Laboratory Manual of Dairy Analysis

Laboratory Manual of Dairy Analysis

— *3rd Revised Edition* —

by

H. DROOP RICHMOND, *F.I.C.*

Chief Analyst to Roots Pure Drug Co. Limited,

Late Analyst to the Aylesbury Dairy Company Limited

BIOTECH

2025

BIOTECH BOOKS

ISBN 978-81-7622-114-6

Published by: **BIOTECH BOOKS®**
4762-63/23, Ansari Road, Darya Ganj,
New Delhi - 110 002
Phone: 09899295259, 011-43003222
E-mail: biotechbooks@yahoo.co.in
Website: www.biotechbooks.in

Typeset at: **Classic Computer Services**
Delhi - 110 035

PRINTED IN INDIA

PREFACE TO THIRD EDITION.

THE text has been revised and augmented by the addition of new methods to bring the work up to date, and the book has been re-indexed.

A description and block of the "Dairy Scale," with which every calculation required in dairy analysis can be made, have been inserted, and to facilitate working out results when the "Dairy Scale" is not available, tables of logarithms and anti-logarithms have been included.

It is hoped that these changes will render the book of increased service.

H. D. R.

PREFACE TO FIRST EDITION.

THIS work is intended to contain working directions for the analysis of milk and dairy-products; the estimation of all constituents of diagnostic value is shortly described in detail, and is in many cases illustrated by photographs of chemists actually carrying out the determination.

A chapter on the application of analysis to the solution of problems usually placed before the chemist is included, and a very short summary of the composition of milk and its products is given.

In the Appendix the composition and preparation of the various solutions is detailed. Tables are given to facilitate the working out of results; these Tables are condensed to occupy one page each, and the saving of time by avoiding the turning over of pages will more than compensate for the slight extra labour due to the condensation.

While not intended to be a complete guide to the analysis of milk, it is hoped that this work will afford assistance to analysts, health officers, dairy students, and those engaged in the supervision of dairies; with this object in view the more simple tests have been described in a manner which will render their working by persons other than chemists possible; it must be remembered, however, that though these methods are easy, they are often fallaciously easy, and lack of chemical training may lead to the making of errors, and the overlooking of important points; no amount of careful following of directions can replace a thorough training in chemical science and manipulation, and though simple tests have a real value as a guide, they have not the reliability of an analysis made by a skilled chemist.

H. D. R.

SYNOPSIS OF CONTENTS.

CHAPTER V.

The Analysis of Butter.

CHAPTER VI.

The Analysis of Cheese.

DAIRY ANALYSIS.

CHAPTER I.

INTRODUCTION.

Milk consists of (1) fat in small globules (Fig. 1) ranging in size from 0·01 mm. in diameter to 0·0016;

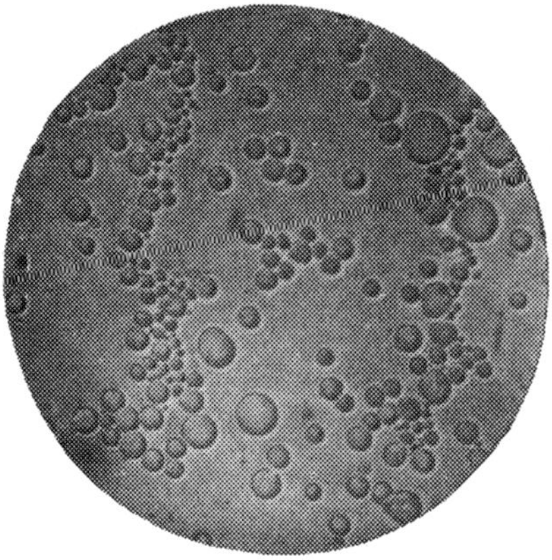

Fig. 1.—Milk (magnified 400 diameters).

(2) milk-sugar and (3) various salts in solution in water; (4) casein, combined with lime and phosphoric acid; and (5) albumin in less perfect solution. There are in addition (6) other compounds in small quantities, including enzymes or natural ferments.

The fat will be treated of in the section on butter; the milk-sugar belongs to the class of carbohydrates and crystallises with 1 OH_2, and is one of the hexabioses. It rotates the plane of polarisation, its specific rotatory power being 52·5° for the crystallised sugar; and reduces solutions of copper salts.

Casein is a protein belonging to the class of the phospho-proteins; it contains carbon, hydrogen, oxygen, nitrogen, sulphur, and phosphorus; in milk it exists as a salt of lime and soda combined with calcium phosphate; acids precipitate the free casein if dilute, while strong acids re-dissolve it. Rennet splits casein up into curd, which is a combination of para-casein with the lime and the calcium phosphate of the casein, the soda being split off, and whey protein which is free from phosphorus.

Albumin is a protein which is distinguished by coagulating on heating to 70° C.; in milk it probably exists as a salt, and this does not coagulate until the milk is acidified. Unaltered albumin is not precipitated by acids.

When micro-organisms act on milk various products are formed; the most important change is the formation of lactic acid from the sugar, which causes milk to become sour, and curdles it by precipitating the casein.

The fat globules are lighter than the aqueous serum, and they tend to rise. *Cream* (Fig. 2) is the upper portion of milk after standing, and differs from milk practically only in that it contains more fat and proportionately less serum. *Skim-milk* is the milk deprived of the bulk of its cream, and if the separation of cream has been performed in a centrifugal separator it is practically free from fat and contains only the aqueous serum. This is termed *separated* or *machine-skimmed milk.*

When cream (or milk) is suitably agitated for some time, the fat globules coalesce to small granules, and these after working together into a nearly homogeneous mass form *butter*. This is chiefly composed of fat, but contains some water and other constituents of milk. The residue is termed *butter-milk*, which does not differ greatly from skim-milk in composition.

By treating milk with rennet, *curd* is separated;

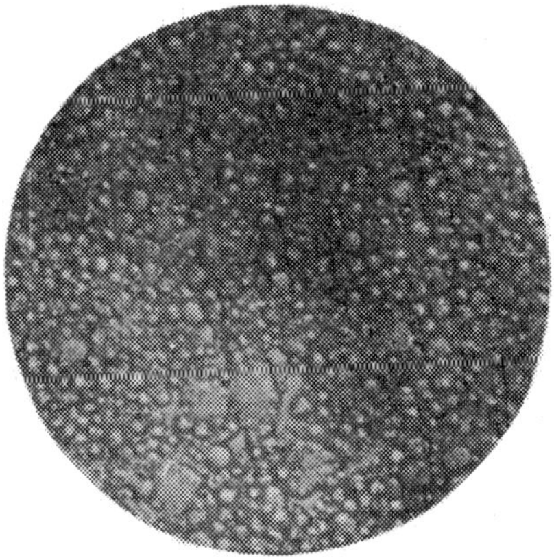

Fig. 2.—Cream (magnified 400 diameters).

this carries down the bulk of the fat, and after pressing, salting, and ripening, partly by the action of micro-organisms and partly by the action of the natural ferments of milk, it is converted into *cheese;* cheese consists essentially of fat, para-casein, and products derived from the latter together with some water and salts.

The following Table gives the average morning and evening milk for each month, and represents the mean percentage composition for 20 years :—

TABLE I.

Month.	Morning Milk.				Evening Milk.			
	Specific Gravity.	Total Solids.	Fat.	Solids not Fat.	Specific Gravity.	Total Solids.	Fat.	Solids not Fat.
January, .	1·0323	12·59	3·65	8·94	1·0321	12·90	3·93	8·97
February,	1·0324	12·52	3·57	8·95	1·0322	12·82	3·87	8·95
March, .	1·0323	12·46	3·52	8·94	1·0320	12·77	3·82	8·95
April, .	1·0322	12·39	3·49	8·90	1·0320	12·69	3·81	8·88
May, .	1·0325	12·28	3·34	8·94	1·0320	12·72	3·78	8·94
June, .	1·0325	12·21	3·30	8·91	1·0320	12·63	3·75	8·88
July, .	1·0320	12·24	3·47	8·77	1·0314	12·55	3·80	8·75
August, .	1·0317	12·32	3·56	8·76	1·0313	12·70	3·95	8·75
September,	1·0320	12·50	3·64	8·86	1·0316	12·89	4·05	8·84
October, .	1·0322	12·66	3·72	8·94	1·0319	13·02	4·10	8·92
November,	1·0323	12·79	3·82	8·97	1·0320	13·09	4·14	8·95
December,	1·0323	12·74	3·77	8·97	1·0320	13·00	4·05	8·95
Average,	1·0322	12·47	3·57	8·90	1·0319	12·81	3·92	8·89

In the Table below is given the average percentage composition of milk and the various products derived from it.

TABLE II.

	Water.	Fat.	Milk-Sugar.	Proteins.	Mineral Matter.
Milk, . .	**87·34**	**3·75**	**4·70**	**3·40**	**0·75**
Separated milk,	90·48	0·12	4·88	3·64	0·78
Thick cream,	39·37	56·09	2·29	1·57	0·38
Thin cream, .	67·50	25·67	3·66	2·60	0·57
Fresh butter,	12·99	85·81	0·37	0·74	0·09
Salt butter, .	13·78	82·97	0·39	0·84	2·02*
Butter-milk, .	90·85	0·46	4·48‡	3·46	0·75
Whey, . .	93·21	0·30	4·99	0·92	0·58
Curd, . .	49·43	27·38	2·04	20·00	1·15
Cream cheese,	30·66	62·99	0·26‡	4·94†	1·15
Soft cheese, .	50·04	27·50	?	18·32†	4·12*
Hard cheese, .	33·89	33·00	1·90‡	27·56†	3·65*
Half-skim cheese. .	37·35	24·61	?	32·40†	5·65*

* Including added salt. † Including products of ripening.
‡ Including lactic acid.

CHAPTER II.

THE ANALYSIS OF MILK.

Preparation of the Sample.—A milk sample conveniently consists of a five-ounce bottle filled nearly full ; if the sample is to be representative of a bulk, the milk (whether in a churn, pail, can, or jug) should invariably be stirred well before the sample is taken in order to distribute the cream which always tends to rise to the surface. Samples taken in a dairy or other place for testing on the premises may, however, be taken in cans, and if the sample is one frequently taken from the same source a distinguishing mark may be stamped on the can.

On receipt of the sample in the laboratory it should invariably be stirred before any portion is withdrawn for analysis ; violent shaking is to be deprecated, as not only is there a tendency to churn the fat but air bubbles, which do not separate immediately, are included, and prevent accurate measurements, especially of specific gravity. In cold weather samples often are frothy, and while they remain cold the air bubbles separate very slowly. Freshly drawn milk is also frothy, and the fat being in the liquid condition has a lower specific gravity than it has after solidification. If the sample is turning sour there is often difficulty in distributing the cream uniformly, and violent shaking may have to be resorted to ; if any of the fat be churned or become churned in this operation, the lumps of churned fat should be removed, dried, and the fat extracted with ether and weighed ; the total weight of milk is ascertained, and the percentage of churned fat calculated. The remainder of the sample is analysed separately.

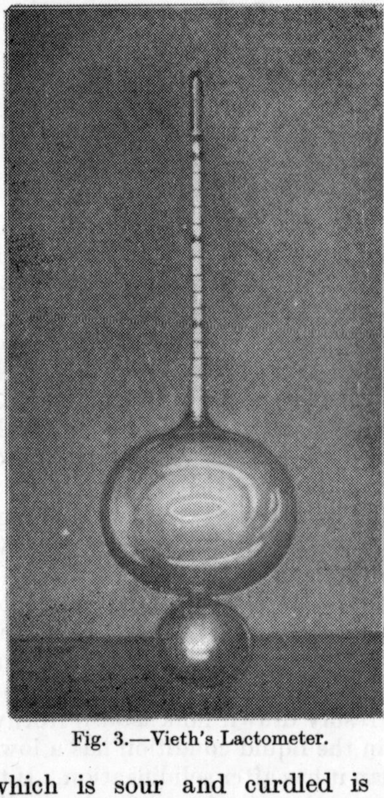

Fig. 3.—Vieth's Lactometer.

Milk which is sour and curdled is mixed by turning the whole sample into a beaker, and whipping with a brush made of fine wires.

Always determine the specific gravity of every sample if possible.

Estimation of Specific Gravity by Lactometer.— Nearly fill a cylindrical vessel of depth such that the lactometer (*see* Fig. 3) will float, and at least $\frac{1}{4}$ inch wider than the lactometer; a glass jar, or tin pot, or even a milk-can serves well. Hold the lactometer at an angle, and lower it cautiously into the milk, taking care that no air bubbles are retained in the space between the upper and lower bulbs; when the upper bulb is partially below the surface, raise to an upright position and immerse the lacto-

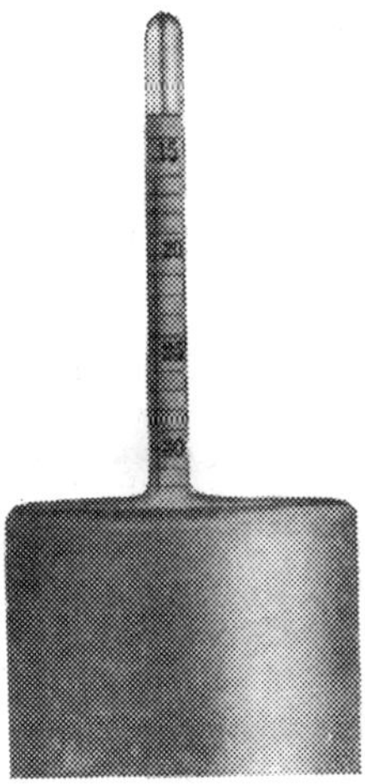

Fig. 4.—Lactometer in Milk.

The jar in the figure has been filled to the brim with milk in order to show clearly the effect of capillary attraction; it is neither necessary nor advisable to do this when testing milk.

meter to the 30° mark, and let it find its own level. When steady read off the point where the surface of the milk cuts the stem; this point is not visible, as the milk is drawn up round the stem by capillary attraction (Fig. 4), and must be estimated mentally. Some little practice is required to do this, and the reading may be obtained by observing the point on the stem to which the milk reaches, and adding a constant amount for the height of the meniscus, usually ½ degree ; thus in the figure the true reading of the lactometer is 32·5°, the apparent reading is

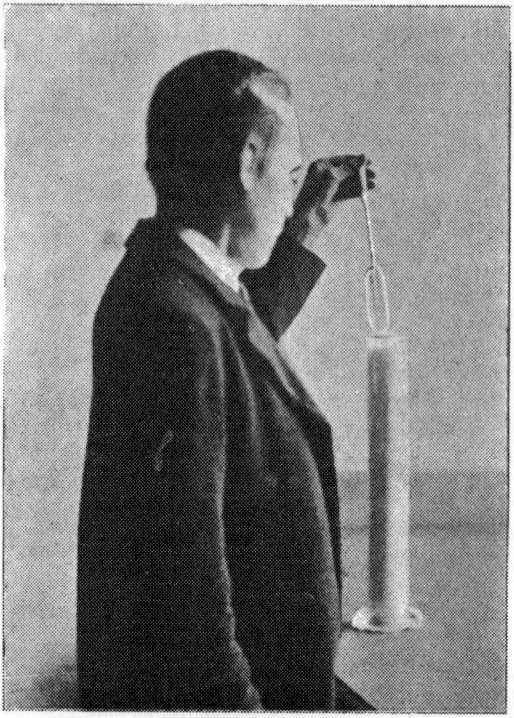

Fig. 5.—Thermo-lactometer.

32°, and with ½ degree added on the true reading is obtained.

Immerse the bulb of a thermometer in the milk, and stir the milk till the reading is constant ; correct the reading to 60° F. by means of the table given on p. 100 or by the milk or dairy scales (p. 44). To use the table, find in the top line the specific gravity (or the nearest figure to the left), and in the left-hand column the temperature; where the lines intersect the corrected specific gravity is given (adding on, if the specific gravity found is not an exact degree, the decimals).

A thermo-lactometer—*i.e.*, a lactometer which contains a thermometer—may be used, and the temperature can then be read off from the upper scale at the same time as the specific gravity (Fig. 5).

Never take a specific gravity reading without also noting the temperature and correcting to 60° F.

The lactometer should be checked by the gravimetric method, to make sure that the scale is correct.

A lactometer does not give accurate results if a film of milk be allowed to dry on it; if a sample has been tested and the lactometer removed, and allowed to stand, it must be washed and dried before being used for another sample.

There is usually little objection, however, to removing a lactometer from one sample and, after draining, placing it at once in another sample; but if the sample is likely to be the subject of legal proceedings or is otherwise important, the lactometer should always be cleaned before testing.

The Gravimetric Estimation of Specific Gravity.—Dry a Sprengel tube (Fig. 6) by washing with distilled water, alcohol, and ether, and placing in the water-oven; aspirate air through while still hot, cool, and weigh. Fill the tube with distilled water, and place it in a vessel containing water at 15·5° C. till the water inside no longer alters in volume; adjust the level of the water accurately to the mark on the wider tube by applying filter-paper cautiously to the narrower end; wipe the outside of the tube dry, and weigh. The difference between the weight of the full and the empty tube gives its capacity in grammes of water at 15·5°.

Empty the water, and rinse the tube several times

Fig. 6.—Sprengel Tube.

with the milk, and then fill it with milk; allow the tube to stand a minute to permit air bubbles to rise, suck or blow these out, and fill the tube completely. Immerse the tube in water at 15·5° C. till the volume ceases to alter, and then wipe dry and weigh as before. The difference between the full and the empty tube gives the weight of milk, and this divided by the capacity in grammes of water at 15·5° gives the specific gravity at 15·5°.

The capacity of the tube once determined remains constant, and it need not therefore be determined every time; the weight of the empty tube should, however, be taken occasionally.

Rise of Specific Gravity of Milk on Standing.—If the milk be freshly drawn or has been recently heated, the fat is in the liquid condition, and its specific gravity is lower than when solid. The solidification of the fat globules takes some time, and twelve to twenty-four hours may elapse before the maximum specific gravity is attained. If the milk be frothy, air bubbles may cause the specific gravity to appear too low.

The maximum specific gravity is taken as the correct figure.

Estimation of Total Solids. Weigh a basin either of platinum, tantalum, fused silica, or porcelain,

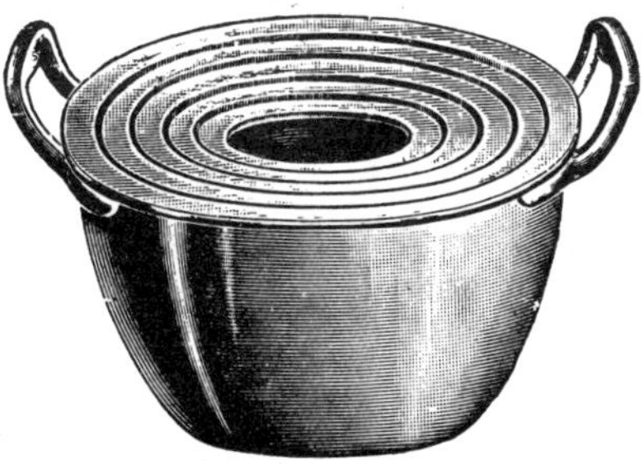

Fig. 7.—Water-bath.

preferably $2\frac{3}{4}$ inches wide and flat-bottomed; pipette in 5 c.c. of milk and weigh again; the weighing should be rapid, but the exactitude of weighing need not be more than 2 mg. Place the basin on a water-bath (Fig. 7), and from time to time break the skin with a needle; when apparently dry place in a water-oven (Fig. 8), and continue the drying for four hours, cool in a desiccator, and weigh; replace in the oven for periods of one hour each, cool and weigh, until the loss in one hour is less than 1 mg. The weight of the residue divided by the weight of milk taken and multiplied by 100 gives the percentage of total solids.

(*Babcock's method*)—The basin may be packed loosely with ignited asbestos, if great accuracy be required; (*Stokes' method*) if speed be wanted, a

Fig. 8.—Water-oven.

few drops of a 10 per cent. solution of acetic acid in alcohol, or (*Revis' method*) an equal volume of acetone, may be added, and the time of drying can then be curtailed.

Estimation of Ash.—The total solids are ignited in a muffle furnace (Fig. 9), which should not be allowed to become hotter than a dull red heat; to facilitate burning, upstanding portions of the ash may be broken down by touching with a platinum wire; when all the carbon is burnt away the basin is cooled in a desiccator and weighed. The ignition may be performed over a Bunsen burner, the point of the flame of which should barely be allowed to touch the bottom of the basin, or an Argand burner may be used; the basin should be covered with a

Fig. 9.—Muffle.

platinum lid. It is very important not to heat the ash to too high a temperature.

Estimation of Soluble and Insoluble Ash.—Fill the basin with hot water and filter through a small ash-free filter; wash with hot water. Place the filter and its contents in the basin, and ignite; the residue corrected for the ash of the filter-paper, if any, is the insoluble ash and the soluble ash can be obtained by difference.

Estimation of Alkalinity, Chlorine, Lime, and Phosphoric Acid.—To the filtrate containing the soluble ash add a drop or two of phenolphthalein solution, and titrate with $\frac{N}{10}$ sulphuric acid (*see* Appendix) till the colour is discharged. Each cubic centimetre of acid indicates 0·0044 gramme CO_2 as carbonates. (*Note.*—The alkalinity is not all due to carbonates, a little phosphate is present.)

To the solution add a few drops of potassium chromate, and titrate with $\frac{N}{10}$ silver nitrate solution (*see* Appendix) till a red colour just appears. Each cubic centimetre of silver solution indicates 0·00355 gramme of chlorine. (*Note.*—A little of the chlorine, about 0·01 per cent., is lost on ignition of the total solids.)

For the estimation of lime and phosphoric acid, it is advisable to take another quantity of milk (10 c.c. or 25 c.c. being preferable); this is dried on the water-bath, ignited, the ash dissolved in a little hydrochloric acid, and the solution boiled; after cooling slightly, ammonia is added drop by drop till a permanent turbidity appears, and hydrochloric acid added in quantity sufficient to remove this, excess being avoided. The solution is brought just to the boiling-point, and a saturated solution of ammonium oxalate added drop by drop, so long as a precipitate appears; the solution is kept hot (in a water-oven) for at least two hours, and filtered through a small ash-free filter; the precipitate is transferred to the filter, washed with hot water, and the filter placed in a tared basin, and ignited over a small flame; when the filter-paper is all burnt away, the precipitate is moistened with a solution of ammonium carbonate, dried, and very gently ignited. The precipitate, now converted into calcium carbonate, is weighed, and the weight of lime found by multiplying by 0·56; it is usually slightly grey, and contains traces of iron, which are small enough to be neglected.

To the filtrate is added 5 or 10 c.c. of magnesia mixture (*see* Appendix), and about one-tenth its volume of strong ammonia; and after stirring well the liquid is allowed to stand at least 12 hours; the precipitate is washed by decantation with dilute ammonia, transferred to a filter, and the washing completed; the filter is placed in a weighed basin, and ignited at first gently, and finally very strongly till white; the residue of magnesium pyrophosphate is weighed, and the amount of phosphoric acid (as P_2O_5) found by multiplying by 0·6396.

For the determination of other mineral constituents, works on mineral analysis should be consulted.

The Estimation of Acidity.—Place 10 c.c. of milk in a white porcelain basin (Fig. 10), add 1 c.c. of phenolphthalein solution (*see* Appendix), and run in from a burette $\frac{N}{10}$ caustic soda, or other alkali solution, strontia being recommended (*see* Appendix), in drops, stirring constantly, till a faint pink colour, equal to that given by adding 1 drop of a 0·01 per cent. solution of rosaniline acetate in 96 per cent. alcohol to 11 c.c. of the same milk, is produced; each $\frac{1}{10}$ c.c. of $\frac{N}{10}$ alkali solution represents 1° acidity, and the degrees multiplied by 0·009 will give the acidity as percentage of lactic acid.

If greater accuracy be required 50 c.c. may be

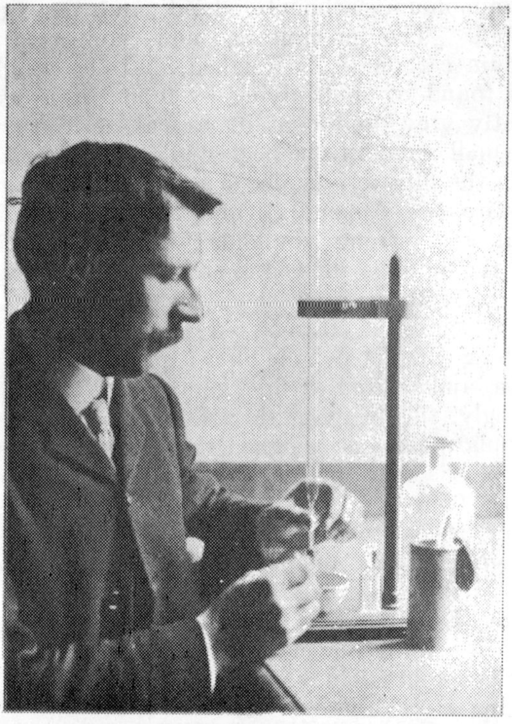

Fig. 10.—Estimation of Acidity.

taken, and 5 c.c. of phenolphthalein solution added; each $\frac{1}{2}$ c.c. will then represent 1° acidity.

The Volumetric Estimation of Fat.—Gerber's method of fat estimation consists in reading the volume of fat brought into the graduated neck of a bottle by centrifuging, after dissolving everything iu the milk but fat by strong sulphuric acid (the essential part of Babcock's method) with the addition of a little amyl alcohol to help the fat to separate (the essential part of Leffmann and Beam's method (Fig. 11)). In chemical principles Gerber's method is nothing but Leffmann and Beam's, and a description of one only differs in details from that of the other; as, however, the

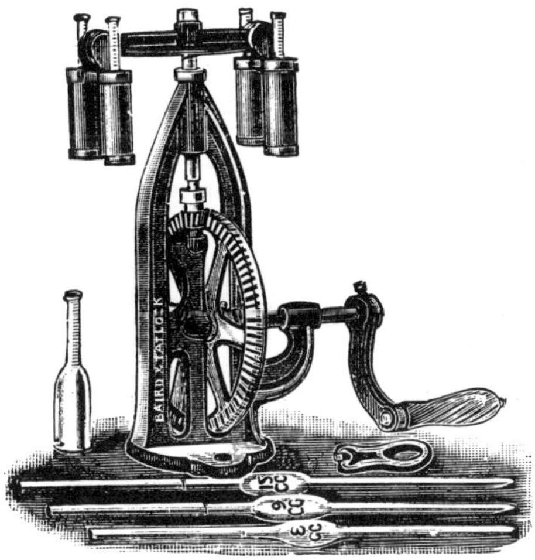

Fig. 11.—Leffmann and Beam Machine.

details of Gerber's method render it more generally suitable, it will be described in preference to its predecessor.

Apparatus.—The essential apparatus consists of:

Acidobutyrometers or Test-bottles.—About 22 c.c. capacity with a long stem expanded to a conical bulb at the top, and graduated in precentages of fat; the bottom is open and contracted to a neck, which is closed by an india-rubber cork when in use. A suitable stand for these is provided.

The stems are made either round, flat, square, or with a magnifying scale; the patterns other than round give a wider and, by many, a more easily read scale.

The test-bottles can now be checked at the National Physical Laboratory, who certify the correctness of the graduated scale.

Measuring Apparatus.—11 c.c. pipettes for measuring the volume of milk taken for the test; 10 c.c.

Fig. 12A.—Gerber Machine.

pipettes for measuring the acid; and 1 c.c. pipettes for the amyl alcohol. Burettes or automatic measuring apparatus may be used.

Centrifuges.—A centrifuge is necessary to bring the fat rapidly into the graduated neck (Fig. 12A). In the smallest size this consists of two long cups hinged on a bar, which is fixed to the top of a vertical spindle running in plain bearings. This is driven by means of a strap round a loose pulley, or by means of a string. The larger machines have a disc carrying four or more cups, provided with a cover, and fixed to the top of a vertical spindle running on ball bearings. The driving is performed either as in the two-bottle machine, by a string wound round the spindle or by a handle. Very large machines are fitted with a steam or water turbine, or an electro-motor (Fig. 12B), and may be provided with heating apparatus to keep the disc warm while running.

Hot-water Tank.—If the disc be not kept warm while running the bottles must be placed in water between 60° and 70° C., before reading, in a small tank.

The Process.—Place a sufficient number of bottles in the stand, open end upwards, and to each add 10 c.c. of sulphuric acid (*see* Appendix); add 11 c.c. of milk to each by means of the 11 c.c. pipette; then add 1 c.c. of amyl alcohol (*see* Appendix).

Fig. 12B.—Gerber Machine.

To measure with a pipette: place the constricted end in the liquid, and draw with the mouth till the level of the liquid is at least one inch above the mark on the upper stem; rapidly remove the mouth, and place the forefinger over the top; the finger must not be wet, though it may with advantage be slightly damp; if this be done with sufficient rapidity, the liquid is above the mark on the upper stem, if not it must be drawn up again. Carefully and slightly raise the finger to allow the liquid to run down slowly to the mark, then stop the flow by pressing the finger on the top; this operation requires some practice, but presents no real difficulty. Keeping the finger pressed down on the top, lift the pipette out of the liquid, and, taking care not to let any drops fall while moving, place the end of the pipette

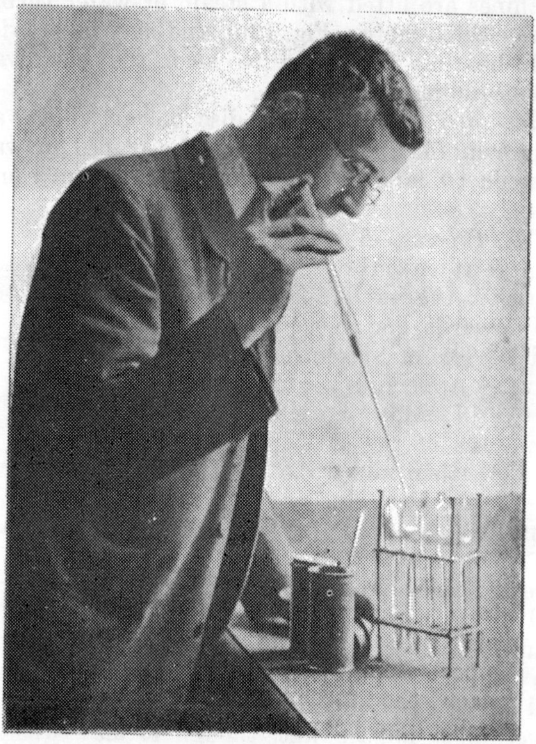

Fig. 13.—Measuring Milk.

inside the neck of the bottle (Fig. 13); hold the pipette slanting, and let the point touch the side of the neck. On lifting up the forefinger the contents will run out down the side of the bottle. Let the pipette drain a few seconds, and remove it, but do not blow out the last drops. It is essential that the milk be measured with great accuracy, but there is less need for exactitude with the acid or amyl alcohol, as slight variations of these do not affect the results; it is as well, however, to cultivate a habit of accuracy.

After measuring the three liquids, which should float in three distinct layers, with little or no browning at the junction of the acid and milk, insert a cork; the bottle should be held with the left hand, and the cork screwed in, not pushed in, with the right; do not exert too much force or the bottle may break. If there be marked browning the milk has either been run in too fast or has not been run down the side of the bottle, and the experiment is best repeated.

Hold the bottle by the stem, and by the cork, keeping a slight pressure on the cork, and mix the contents by shaking; when all the white particles of curd have disappeared, but not before, invert the bottle to allow the acid to run out of the neck, and then turn it upright; repeat this several times till the acid that was in the neck has completely mixed. The bottle is now almost too hot to hold, from the heat developed by the action of the acid, and its colour is brown; place it in the centrifuge. Treat the other bottles in the same way, and arrange them in order in the centrifuge. It is well to mark each cup with a number to avoid errors.

The shaking of all the bottles may be done at once by means of the author's stand (Fig. 14), in which the bottles are held by being pushed into slits in the india-rubber plate; the hand should be placed over all the corks to prevent them from coming out, and the contents of the bottles from being spilt. Other convenient stands are also made.

If a number of samples insufficient to fill the disc is being tested, care should be taken to place them symmetrically so as to preserve the balance of the machine ; a bottle filled with a mixture of equal parts of acid and water may be kept to make up

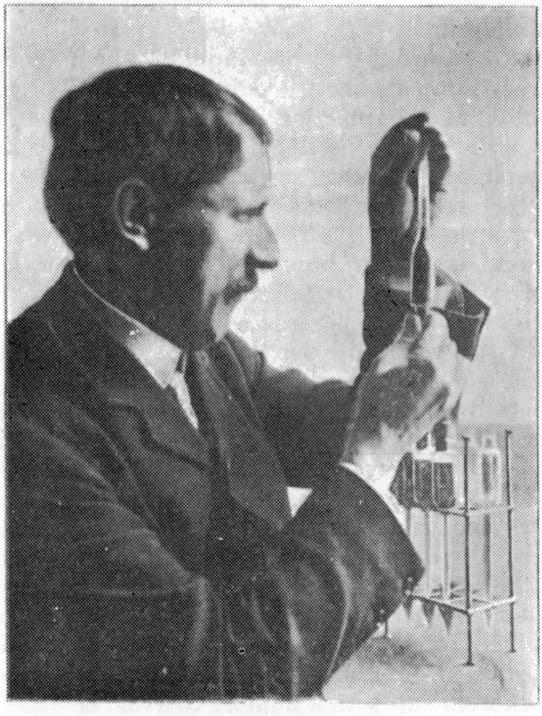

Fig. 14.—Reading Bottles.

an even number of bottles should an odd number of samples be tested.

Screw on the cover, and rotate the machine at about 1,000 revolutions per minute for three or four minutes ; modern centrifuges are usually turned with a handle ; if there be a plain spindle, with a projecting piece, place the eye of the cord on the projection and turn the machine round counter-clock-wise till the cord is wound round the spindle ; take the handle in the hand, pull hard, and the disc will spin ; usually this must be done twice. If there is a strap, place this half round the pulley, and pull with the right hand in a somewhat downward direction, keeping the strap taut with the left hand, and continue giving sharp pulls till the disc spins rapidly ; the pull must be downward as well as forward, or the pulley will not engage the spindle ; when the speed slackens, a few further pulls are necessary. If a string with two handles is provided, this is wound once completely round the spindle (or the pulley on the spindle) ; take a handle in each hand, and pull with the right hand, keeping the string taut with the left hand, and at the end of the pull continue the motion of the left hand to loosen the string round the spindle ; pull back with the left hand keeping the string very loose, and then repeat the stroke. This method of driving requires care, but is, when learnt, a most satisfactory way of spinning the disc ; when a sufficient speed is attained the string is allowed to hang loosely, the handles being placed on the bench at each side. If the string be not properly loosened at the end of the stroke, or on the return stroke, it winds up, and the handles must then be immediately dropped, or the hands may receive a nasty blow. Unbleached blind-cord is a suitable material for the string, and a good supply should be kept, as the string wears.

During the running the disc may be kept warm by means of a Bunsen burner or a spirit-lamp placed underneath near the edge, with a flame so adjusted that it just touches the disc.

The disc should be stopped gently, not suddenly, and the cover unscrewed. Do not take hold of the boss, in the centre of the cover, when the machine is running at high speed.

If the disc be not kept warm, place the bottles, after stopping the machine and removing the cover, in water at 60° to 70° C.; the small tank provided with the apparatus is nearly filled with water at the required temperature, and a spirit-lamp or Bunsen burner used to keep it warm.

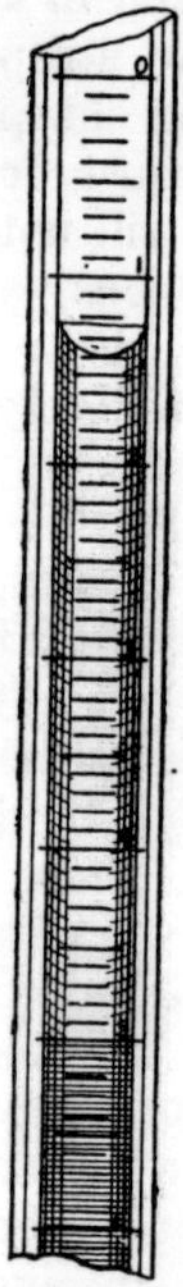

Fig. 15. Gerber Bottle.

After one minute the bottles may be read; hold the bottle (Fig. 14) by the top with the left hand and by the cork with the right, at a level with the eye; the position of the bottle should be as nearly vertical as possible. By gently moving the cork, the lower level of the fat column is adjusted to one of the longer lines indicating percentages of fat, and the point on the scale where the lowest part of the curved upper limit stands is read off; a convenient method of reading is to count first the number of whole percentages of fat (indicated by the longer lines), and then the number of small lines above this, each small line indicating 0·1 per cent. of fat. Fig. 15 shows a sample reading 3·6 per cent. of fat; it is quite easy to read to half a small division or 0·05 per cent.

Every bottle should be read twice to make sure that there is no error in the first reading.

The following are sources of error:—

(*a*) Faulty graduation of the bottle; this is rare.

(*b*) Chemicals not equal to specification; it is important that the specification for these be adhered to strictly.

(*c*) Insufficient mixing of the milk, acid, and amyl alcohol; this is indicated by the fat being cloudy, and obstinately refusing to become clear.

(*d*) Mixing of the milk and acid before addition of the amyl alcohol; a brown colour of the fat is usually found here.

(*e*) Allowing a portion of the fat to remain in the little conical bulb; if the fat be too low down on the scale for convenient reading, and the cork be not pushed in carefully, the fat is liable to jump up, as it is raised; if the fat is too high, it may partially occupy the conical bulb; in each case the fat must be allowed to run down before reading.

When the corks have been used for some time, they do not fit the necks of the bottles so well as when new, and are liable to slip out, and spill the acid mixture; the times when they are most liable to come out are, when shaking the bottle, when removing it from the disc, when removing it from the warm water, and when pulling down the cork to adjust the fat layer. As acid is detrimental to the clothes, always keep a bottle of ammonia handy, and soak the acid-bespattered garment liberally with this, should any be spilt; if acid be spilt on the hands or face, a copious stream of cold water is the remedy, but do not use ammonia; even strong sulphuric acid spilt on the flesh rarely does harm if washed off at once, and plenty of water be used.

When the bottles have been read, turn them cork upwards in the stand, remove the corks, and empty the contents of the bottle into a convenient pot, not down a sink. The pot when full may be emptied down a well-flushed drain; rinse the bottles two or three times in hot water and leave them to drain open end downwards; an occasional clean with a brush when the inside becomes dingy is necessary. The corks should be put in a basin and washed several times with hot water and allowed to dry.

Care of the Machine.—The bearings require lubrication; an oil-can and a supply of suitable oil is provided with the machine. If the ball-bearings wear loose, they can be adjusted by means of a collar in the top bearing. If a bottle break in the machine, remove the cup and wash it carefully, and should acid be spilt on the disc, wash this too.

The machine should be clamped and screwed to a firm bench or table, preferably close to a leg.

If separated milk is being examined a special bottle with narrow neck should be used, and the time of centrifuging increased.

Cream cannot be tested direct, as the scale does not go beyond 9 per cent., or in some bottles 6 or 7 per cent., and a special bottle may be used. But cream can be tested in the ordinary bottles by the following method:—If a thin cream (containing less than 32 per cent. fat) is to be examined, use a 3 c.c. pipette in place of the 11 c.c. pipette; fill the bottle with acid as before, and then add 8·2 c.c. of water from a pipette graduated in $\frac{1}{10}$ c.c.; fill the 3 c.c. pipette with cream, adjusting the upper level very accurately to the mark, and wipe the outside of the stem; hold the pipette vertically over the centre of the opening and allow the cream to run directly into the water and blow out the last drops; finally, add the amyl alcohol, and proceed as described for milk.

For thicker creams, a small balance weighing to

0·05 gramme, and a pair or pairs of accurately balanced tin pots are necessary; place sufficient cream in one pot to fill it nearly half full, and put this on one pan of the balance; place the other pot on the other pan and pour in separated milk or water till exactly balanced. Mix the contents of the two pots by pouring backwards and forward several times, and measure out the mixture as directed for thin cream.

Use the following Table for the calculation of results; the first column is the reading observed; if thin cream was tested, its percentage of fat is found in the column headed "Undiluted"; if thick cream was examined, the column headed "Diluted" will give the percentage of fat.

TABLE III.

FOR CALCULATING FAT IN CREAM.

The Table should be checked by a gravimetric method, and may require a slight correction added or subtracted which may vary with each pipette.

Degrees.	Undiluted.	Diluted.	Degrees.	Undiluted.	Diluted.
8·5	33·2	66·2	6·7	25·9	51·6
8·4	32·8	65·4	6·6	25·5	50·8
8·3	32·4	64·6	6·5	25·1	50·0
8·2	32·0	63·8	6·4	24·7	49·2
8·1	31·6	62·9	6·3	24·3	48·4
8·0	31·2	62·1	6·2	23·9	47·6
7·9	30·7	61·3	6·1	23·5	46·8
7·8	30·3	60·5	6·0	23·1	46·1
7·7	29·9	59·7	5·9	22·7	45·3
7·6	29·5	58·9	5·8	22·3	44·5
7·5	29·1	58·1	5·7	21·9	43·7
7·4	28·8	57·3	5·6	21·5	42·9
7·3	28·3	56·4	5·5	21·1	42·1
7·2	27·9	55·6	5·4	20·7	41·3
7·1	27·5	54·8	5·3	20·3	40·5
7·0	27·1	54·0	5·2	19·9	39·7
6·9	26·7	53·2	5·1	19·5	38·9
6·8	26·3	52·4	5·0	19·1	38·1

In the " Sinacid," " Sal," and " Neusal " methods an alkaline solution is substituted for the acid of the Gerber process, and iso-butyl alcohol for the amyl alcohol. Gerber bottles are used for these methods; 11 c.c. of alkaline solution, 10 c.c. of milk, and 0·6 c.c. of iso-butyl alcohol (which is usually coloured by a dye) are measured into the tubes. The mixing and centrifuging are done as in the Gerber method, but the temperature of the water-bath is 45° C. The coloured alcohol passes into the fat, and colours the layer, which is consequently easy to read.

While the substitution of the alkaline solution avoids the objections to a strong acid, this method has the drawback that the corks become slippery and tend to come out of the bottles.

Gravimetric Estimation of Fat.—Of the numerous methods for the estimation of fat, the Storch, the Ritthausen, the Werner-Schmid, and the Gottlieb methods all present advantages.

The Storch Method.—Place three or four grammes of ignited kieselguhr in a basin, and pipette 10 c.c. of milk in such a manner that it is all absorbed by the kieselguhr; dry on the water-bath with occasional stirring to break up lumps, grind fine, and transfer to a fat-free thimble, rinsing out the basin with fresh quantities of kieselguhr; place the thimble in a Soxhlet extractor, rinse the basin, pestle, etc., with ether, and pour this into the thimble; extract the kieselguhr with ether.

The Ritthausen Method.—*See* Estimation of Proteins, p. 35.

The extraction in these methods is performed by attaching a weighed flask to the bottom of the extractor to receive the ether containing the fat, and connecting the extractor to an upright condenser; the flask is immersed in water kept warm by a small flame, and the ether continually distils up, is condensed, and runs back into the extractor, from which it siphons back into the flask when the extractor is full.

The kieselguhr should be extracted about four

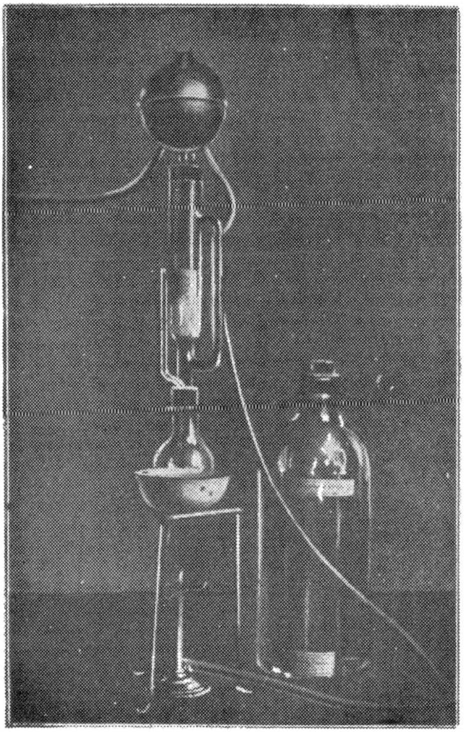

Fig. 16.—Soxhlet Extractor.

hours, and casein about one and a half hours, after which the ether is distilled from the flask, and the fat freed from ether by placing in a water-oven for twenty minutes, blowing into and rotating the flask every five minutes. After cooling for a quarter of an hour, the flask is weighed, and the increase of weight represents the fat.

It is advisable to return the flask to the water-oven for a second period, and re-weigh to make sure that all the solvent has been removed. The fat may be melted and dissolved in petroleum ether which is carefully decanted from any residue ; by successive treatments with small quantities of petroleum ether all the fat may be removed, and the flask, with any small particles of kieselguhr that may have passed into it, re-weighed.

Fig. 17. Stokes Tube.

The Werner-Schmid Method.—Pipette 10 c.c. of milk into a Stokes tube (Fig. 17), add 10 c.c. of strong hydrochloric acid, and heat over a flame *with constant shaking*, till the fat, on standing a short time, collects in a clear layer on the surface ; cool the contents of the tube, and add 30 c.c. of ether, cork the tube, and shake well ; allow the tube to stand till the ether has separated in a clear layer ; remove as much ether as possible, preferably by means of washbottle tubes, to a weighed flask ; add about 20 c.c. more ether, shake well, allow the ether to settle clear, and remove as before, and again add about 20 c.c. of ether, shake, allow to settle, and remove the clear layer. Distil off the ether, dry, and weigh as above.

Gottlieb's Method.—Place 5 c.c. of milk in a stoppered tube, add successively 0·5 c.c. of ammonia, 5 c.c. of alcohol, $12\frac{1}{2}$ c.c. of ether, and mix thoroughly, finally add $12\frac{1}{2}$ c.c. of petroleum ether ; mix well, and allow to stand ; when a clear layer has separated, mix again, allow to separate, mix once more, and allow the clear layer to separate completely.

Remove the ethereal layer as completely as possible by washbottle tubes to a flask, and add and remove three successive portions of a mixture of equal parts of ether and petroleum ether, the recovered solvent serving well. Evaporate the solvent, dry, and weigh as above. Extract with petroleum ether and weigh the empty flask.

In all methods the percentage of fat is calculated by multiplying the weight by 100, and dividing by the weight of milk taken.

Estimation of Milk-Sugar.—This may be estimated either polarimetrically, or gravimetrically; the results in either case are expressed as anhydrous sugar.

Polarimetric Estimation.—50 c.c. of milk are measured into a dry flask, and a quantity of water equal in cubic centimetres to the sum of

(*a*) The degrees of gravity divided by 20.
(*b*) The percentage of fat divided by 1·8.
(*c*) A quantity to convert scale readings into percentages of anhydrous sugar; if the scale be in angular degrees and a 200 mm. tube is used, this is 5·43 c.c. (or 5 c.c. with a 198·4 mm. tube).

1·5 c.c. of Wiley's acid mercuric nitrate solution (*see* Appendix) is added, and the whole well mixed by violent shaking. The solution is poured on a dry filter, and a polarimeter tube filled with the clear filtrate.

As an example: the milk has a sp. gr. of 1·032, the degrees of gravity are 32·0, and (*a*) is $\frac{32 \cdot 0}{20} = 1 \cdot 60$ c.c.; the fat is 3·60 and (*b*) is $\frac{3 \cdot 6}{1 \cdot 8} = 2 \cdot 00$ c.c.; if an instrument graduated in angular degrees and a 200 mm. tube are used (*c*) is 5·43. The water added is $1 \cdot 60 + 2 \cdot 00 + 5 \cdot 43 = 9 \cdot 03$ c.c.

The reading is made by placing the tube in the instrument (Fig. 18), focusing the eye-piece on the half-shadow plate when the analyser is so turned that one side is darker than the other, and adjusting the analyser till both sides are equal in intensity; the scale is then read by means of the vernier provided. Several readings should be made, the adjustment to equality being made from either

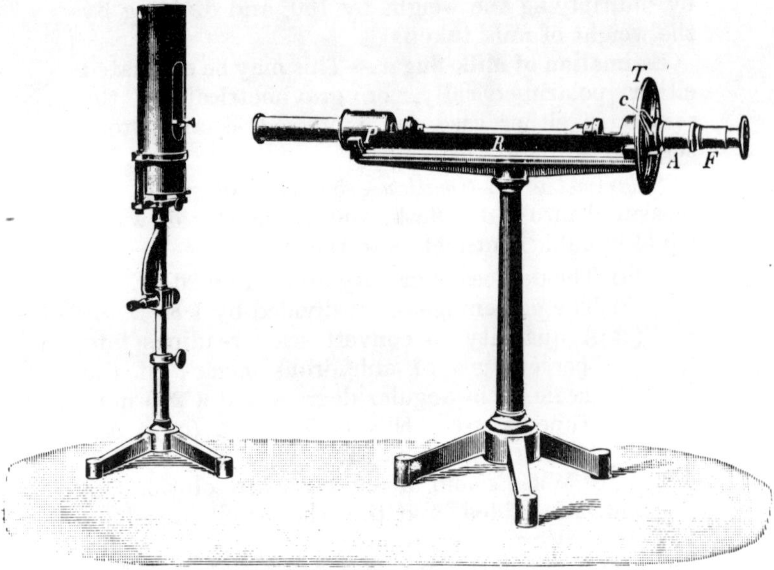

Fig. 18.—Polariscope.

side alternately, and the mean of the readings taken as the correct reading. A blank estimation —*i.e.*, one with a tube filled with distilled water—should always be made, and the reading if to the right subtracted from, or if to the left added to, the reading of the sample. The corrected reading of the scale gives the percentage by weight of anhydrous milk-sugar.

Mercuric nitrate does not remove the proteins quite completely; for greater accuracy add to the filtrate $\frac{1}{20}$ of its volume each of phospho-tungstic acid solution and of dilute (1 : 1) sulphuric acid solution, filter and polarise. Multiply the reading by 1·1. The error with whole milk is, however, very small and hardly exceeds the experimental error of reading.

Gravimetric Estimation.—About 10 grammes of milk are placed in a 100 c.c. flask with 60 to 70 c.c. water, 5 c.c. of Fehling's copper sulphate solution added (*see* Appendix), and the solution neutralised with caustic soda; the liquid is made up to 100 c.c., and the contents of the flask, after mixing, filtered through a dry filter; 50 c.c. of the filtrate are placed in a beaker, and a mixture of 30 c.c. each of Fehling's copper sulphate and alkaline tartrate solutions (*see* Appendix) added; the beaker is heated by a flame of such size that the liquid boils in about four minutes, and it is kept boiling for exactly six minutes. A tube of hard glass, about 1 cm. in internal diameter and 10 cm. long, with one end drawn out, and plugged with a fairly tight wad of asbestos, is ignited, cooled, and weighed; the drawn-out end is inserted in a hole in the cork of a vessel from which the air can be exhausted (*e.g.*, by a filter pump), a small funnel is placed in the wider end, and the liquid carefully poured off from the red precipitate of cuprous oxide; the precipitate is washed several times by decantation with hot, well-boiled water, and is then washed into the tube, and the last traces removed from the sides of the beaker by rubbing with a policeman, and transferred to the tube; the precipitate in the tube is well washed with hot water, and finally with alcohol, and dried. The narrow end of the tube is connected with an apparatus from which hydrogen can be evolved, and after the stream has passed for a few minutes the part of the tube containing the precipitate is heated gently by a small flame, till the cuprous oxide is reduced to copper; the hydrogen is passed till the tube is cool, and it is then disconnected and weighed; the increase in weight gives the quantity of copper, from which the quantity of milk-sugar can be found by the table below; this multiplied by 200, and divided by the weight of milk taken, gives the percentage of milk-sugar.

TABLE IV.

FOR CALCULATING WEIGHT OF MILK-SUGAR FROM COPPER REDUCED.

(All weights are in milligrammes.)

Copper	Milk-Sugar	Copper	Milk-Sugar	Copper	Milk-Sugar
120	86·4	215	158·2	310	232·2
125	90·1	220	161·9	315	236·1
130	93·8	225	165·7	320	240·0
135	97·6	230	169·4	325	243·9
140	101·3	235	173·1	330	247·7
145	105·1	240	176·9	335	251·6
150	108·8	245	180·8	340	255·7
155	112·6	250	184·8	345	259·8
160	116·4	255	188·7	350	263·9
165	120·2	260	192·5	355	268·0
170	123·9	265	196·4	360	272·1
175	127·8	270	200·3	365	276·2
180	131·6	275	204·3	370	280·5
185	135·4	280	208·3	375	284·8
190	139·3	285	212·3	380	289·1
195	143·1	290	216·3	385	293·4
200	146·9	295	220·3	390	297·7
205	150·7	300	224·4	295	302·0
210	154·5	305	228·3	400	306·3

Volumetric Estimation.—See p. 109.

Estimation of Proteins.—These may be estimated together by Ritthausen's method, or the casein and albumin may be separated; indirect estimations may be made from the nitrogen by Kjeldahl's method, from the "aldehyde figure," or from the organic phosphorus and sulphur.

The Ritthausen Method.—Pipette 10 c.c. of milk into a beaker, and add 100 c.c. of hot water; add 5 c.c. of Fehling's copper sulphate solution (*see*

Fig. 19.—Washing Precipitate.

Appendix), and neutralise with caustic soda solution; collect the precipitate either in a weighed Gooch crucible, or on tared filter-paper; remove the precipitate completely from the beaker by means of a policeman, and wash well (Fig. 19). Dry in the water-oven, and extract the fat with ether, preferably in a Soxhlet extractor, and dry again till the weight is constant. Ignite the precipitate in the Gooch crucible, or place the filter containing the precipitate in a weighed basin and ignite it (if tared filters are used the tare should also be ignited in a weighed basin); subtract the weight of the ash (corrected if necessary for the ash of the tare) from the weight of the precipitate, and the difference will give the weight of the proteins.

Estimation of Casein and Albumin.—Pipette 10 c.c. of milk into a beaker, add 90 c.c. of water at 42°-43° C. and 1·5 c.c. of a 10 per cent. solution of acetic acid; stir well, and collect and weigh the precipitate as above. (*Note.*—This estimation or that given above may be combined with a fat estimation.) The precipitate is in this case casein.

Raise the filtrate to boiling, and keep for fifteen minutes on the water-bath; collect the precipitate as above, and weigh after drying; as fat and ash are absent the extraction and ignition may be omitted. The precipitate consists of albumin.

Estimation of Nitrogen by Kjeldahl's Method.—5 c.c., or 10 c.c., of milk are pipetted into a long-necked hard glass flask; 20 c.c. pure sulphuric acid and a drop of mercury are added; a long-stemmed bulb is placed in the neck; the flask is supported in an inclined position, and it is heated by a small flame; at first water is driven off, next a considerable amount of frothing takes place, and when this has subsided the flame may be turned up to such a height that the sulphuric acid distils up to and is condensed in the neck; 10 grammes of acid potassium sulphate or 8 grammes of exsiccated sodium sulphate may be added after the frothing has subsided, but this addition is hardly worth while with milk, as the operation is fast enough without it; the heating is continued till the liquid in the flask is quite colourless. After cooling, the acid liquid is diluted with about 100 c.c. of water, and poured into a flask of at least 1,000 c.c. capacity (preferably of copper (Fig. 20)), and the

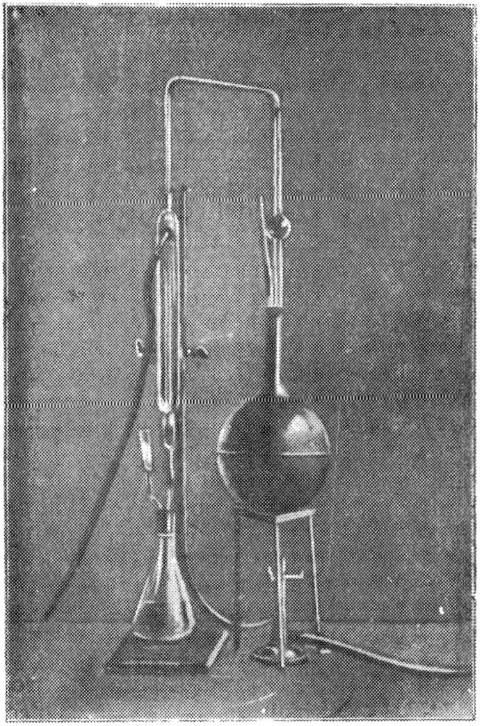

Fig. 20.—Kjeldahl Apparatus.

hard glass flask rinsed out with further quantities of water; the large flask is furnished with a cork carrying a tube which is connected to a condenser, and a tube allowing additions to be made to the liquid. 25 c.c. (50 c.c. if 10 c.c. of milk were taken) of $\frac{N}{10}$ acid, or an amount calculated to be slightly more than sufficient from the aldehyde figure (p. 40), are accurately measured into a flask, which is placed so that the end of the condenser dips into the liquid. Through the tube in the cork 100 c.c. of a solution of caustic soda containing 300 grammes per litre are poured in, followed by 10 c.c. of a 10 per cent. solution of sodium sulphide; the tube is closed, and the contents of the flask mixed by shaking; a flame is placed under the flask, and the ammonia formed by the action of the sulphuric acid on the proteins and liberated by the alkali, distilled over; the distillation takes about half an hour, a volume of approximately 200 c.c. being collected. A few drops of methyl red or cochineal are added to the distillate, and the excess of acid titrated with $\frac{N}{10}$ alkali (*see* Appendix). A blank experiment—*i.e.*, one without the milk—should be performed, and the difference between the volume of alkali used in the blank and the experiment in cubic centimetres multiplied by 0·0014 will give the weight of nitrogen; this multiplied by 100 and divided by the weight of the 5 c.c. or 10 c.c. taken will give the precentage of nitrogen; the nitrogen multiplied by 6·39 may be taken as proteins in the milk.

Estimation of Casein and Albumin Nitrogen.—To 10 c.c. of milk add 20 c.c. of a saturated solution of magnesium sulphate, and crystals of the salt so long as they are dissolved; allow to stand for some time, and filter off the precipitated casein, and wash this several times with the saturated solution of magnesium sulphate (this is a slow process); place the filter and precipitate in a long-necked, hard glass flask, add 30 c.c. of pure sulphuric acid and a drop of mercury, and proceed as above; 150 c.c. of soda solution must be added before distillation, and the distillate should be collected in 35 or 40 c.c. of $\frac{N}{10}$ acid.

The filtrate is diluted, and the albumin precipitated by tannin; the precipitate is collected on a filter, and the nitrogen estimated as above.

The casein nitrogen and albumin nitrogen may be multiplied by 6·39 to give the amounts of casein and albumin respectively; the sum of the two is usually a little below the total nitrogen.

Indirect Estimation from Organic Phosphorus and Sulphur.—Estimate the phosphoric acid in the ash of the milk (*see* p. 15), preferably using 25 grammes of milk; evaporate 25 c.c. of the filtrate obtained by adding Wiley's mercuric nitrate to milk for the polarimetric estimation of milk-sugar (*see* p. 31), ignite and estimate the phosphoric acid in the ash of this; multiply the weight of P_2O_5 by 4·42, and this will give the percentage of mineral phosphoric acid in the milk; this subtracted from the percentages of total phosphoric acid will give the organic phosphoric acid, and the casein can be calculated from this by multiplying by 50·8.

Take 25 c.c. of milk, add 10 c.c. of strong nitric acid and 1 gramme of sodium carbonate, evaporate to dryness, and add 2 or 3 c.c. more nitric acid, and again evaporate; ignite to a fairly white ash, and take up the residue with dilute hydrochloric acid; evaporate to dryness, and boil the residue again with dilute hydrochloric acid, and filter. Just raise the filtrate to boiling, and add barium chloride solution drop by drop so long as a precipitate is produced. Allow the solution to stand 24 hours to complete the precipitation, the first two hours preferably in the water-oven, and collect the precipitate on a small filter, wash well with hot water, and place filter and precipitate still wet in a weighed platinum dish; ignite over a small flame till the paper is burnt, and then more strongly; the weight of the precipitate (corrected for the ash of the filter) multiplied by 13·7 and divided by the weight of milk taken will give the percentage of sulphur in the milk. For each 71 parts of organic P_2O_5, 32 parts of sulphur belong to the casein; the remainder of the sulphur multiplied by 58·5 will give the albumin.

Aldehyde Figure.—This determination may be combined with an acidity estimation; to 10 or 11 c.c. of milk at least 1 c.c. of 0·5 per cent. phenolphthalein solution is added, and the milk neutralised with $\frac{N}{11}$ (approximately) strontia solution; to the faintly pink liquid 2 c.c. or more of 40 per cent. formaldehyde solution is added, and the titration continued till the same degree of pink colour appears. After deducting the acidity of the formaldehyde solution, the latter titration represents the aldehyde figure. The aldehyde figure is obtained by multiplying the number of c.c. used by the strength of the solution (*see* p. 107) and by 1,000, and dividing by the volume taken. Proteins may be deduced from this figure by multiplying by 0·170; this factor is only applicable to fresh cow's milk when determined by strontia solution.

The relation between nitrogen, proteins, and aldehyde figure is given on the dairy scale.

Catalase.—For the estimation, the measuring cylinder is filled with water through the opening *d*, the cover of which is then screwed down. The opening *b* (for cleaning the tube) is closed with a rubber cork, and the chamber *A* is charged, through the opening *c*, with 15 c.c. of the milk and 5 c.c.

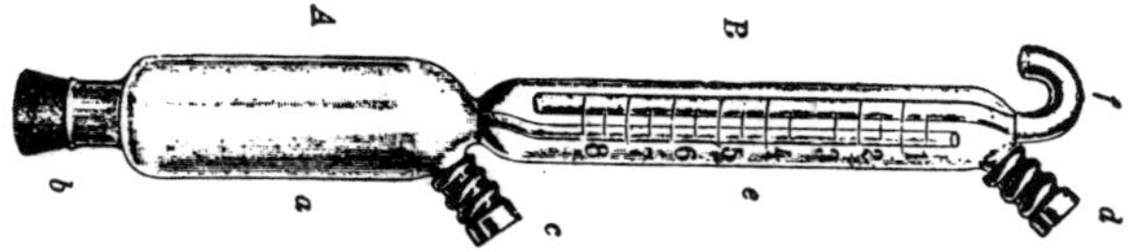

Fig. 21.—Lobeck's Catalase Tube.

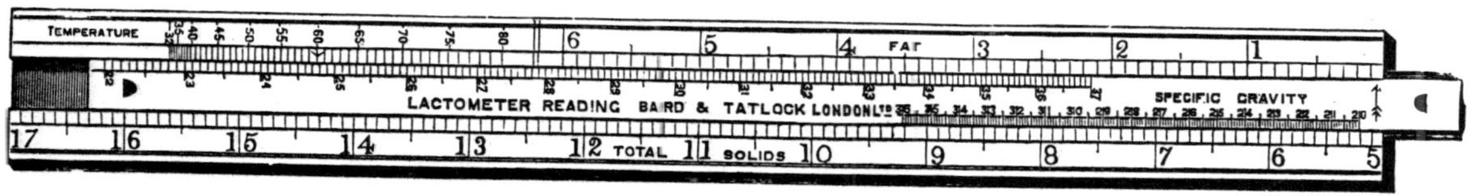

Fig. 22.—Richmond's Milk Scale.

of 1 per cent. hydrogen peroxide solution (or 9 c.c. of milk and 3 c.c. of hydrogen peroxide). The tube is now held at *f* and *d* and shaken with a pendulum motion, and the cover to *c* rapidly screwed down. The tube is then placed in water at 25° C. up to the level of *c*, shaken from time to time, and the volume of the liberated gas which rises into the measuring chamber *B* read off after two hours.

The Relation between Fat, Solids not Fat, and Specific Gravity.—As the solids not fat of milk are heavier than water, and the fat is lighter, and as, moreover, the extent to which each of these is heavier or lighter respectively is practically constant, it is evident that by means of an appropriate formula any one of the three can be approximately calculated from the other two.

A simple formula which gives a very good approximation is :

$$S = \frac{G}{4} + \frac{F}{5} + 0{\cdot}14, \qquad (1)$$

where S = Solids not Fat per cent. by weight ;
G = degrees of gravity ; and
F = Fat per cent. by weight.

If it be desired to calculate directly the Total Solids (T), which consists of the sum of the fat and solids not fat, we may write

$$T = S + F = \frac{G}{4} + 1{\cdot}2\,F + 0{\cdot}14, \qquad (2)$$

As the solids not fat are not generally estimated directly, but are obtained by subtracting the fat from the total solids, the second formula is more useful when the fat is to be calculated from the specific gravity and total solids, and it may be usefully converted to the form

$$F = 0{\cdot}833\,T - 0{\cdot}208\,G - 0{\cdot}12, \quad . \quad (3)$$

or perhaps more simply

$$1{\cdot}2\,F = T - \frac{G}{4} - 0{\cdot}14, \quad . \quad . \quad (4)$$

Tables for the easy calculation of solids not fat

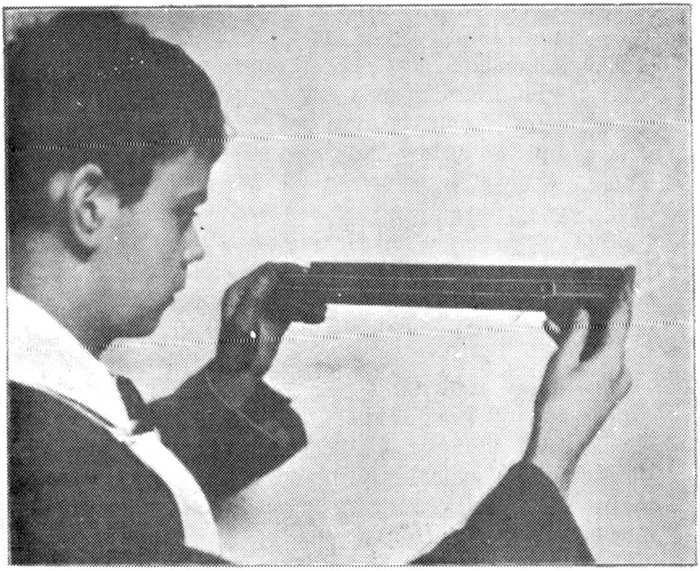

Fig. 23.—Using Milk Scale.

from fat and specific gravity, and fat from total solids and specific gravity are given at the end of the book (pp. 101, 102). As the method is only approximate, the figures are calculated only to the nearest 0·05 per cent. To use these Tables, find in the upmost horizontal column the specific gravity, and in the vertical column the fat or the total solids ; in the space where the columns intersect the figure required is found. If the last figure in the specific gravity is 9, 0, or 1, or 4, 5, or 6, use the figure in the column corresponding to the nearest 0 or 5 ; if it is 2 or 7, use the column corresponding to 0 or 5, but add in Table XII. 0·05, and subtract 0·05 in Table XIII. ; if it is 3 or 8, use the column corresponding to 5 or 0, and subtract 0·05 in Table XII. and add 0·05 in Table XIII. Read directions for use of each Table.

The calculation may also be made with the milk scale, which consists of a slide rule (Figs. 22, 23) ; on one side is marked total solids (1 inch = 1 per cent.), on the other the fat (1·2 inches = 1 per cent.), on the slide is marked specific gravity ($\frac{1}{4}$ inch = 1 degree), and an arrow is placed 0·14 inch from the end of the scale. If the arrow be placed against the fat, the specific gravity lies against the total solids ; *vice versâ*, if the specific gravity be placed against the total solids the arrow will point to the fat.

The dairy scale (Fig. 24), which is a greatly improved milk scale, may be used to calculate either total solids or solids not fat ; if the arrow on the upper scale be adjusted to the percentage of fat the total solids can be read off against the specific gravity, and by adjusting the arrow on the slide to the "fat for S.N.F." scale, the solids not fat may be read off against the specific gravity, the reading in each case being facilitated by the cursor provided.

Every calculation required in dairy analysis can be made with the dairy scale. It includes a slide rule.

On another part of either rule is a scale for correcting the specific gravity taken at any temperature to 60° F. On the rule is a scale of degrees

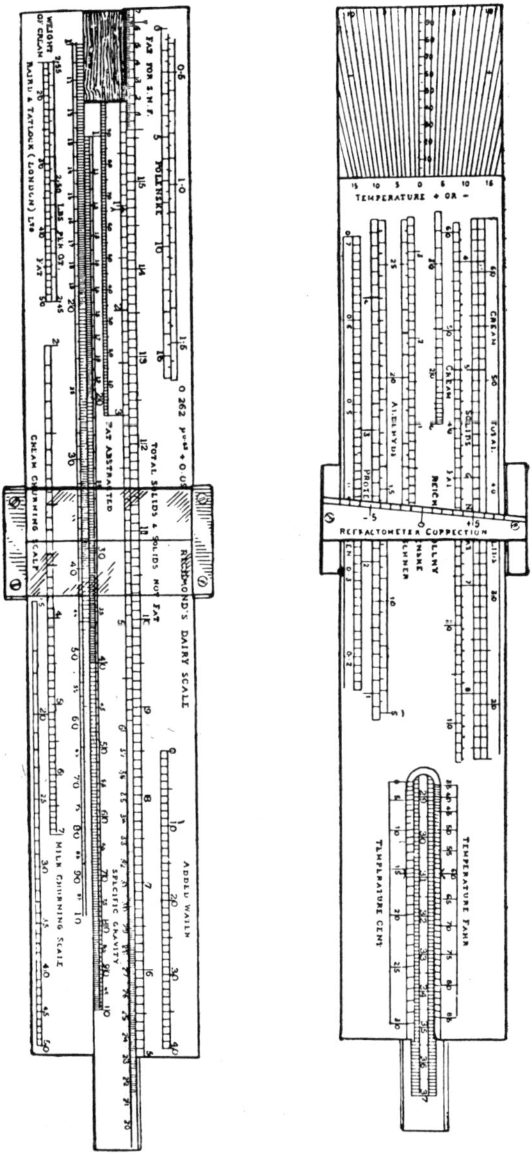

Front View. Back View.

Fig. 24.—Dairy Scale.

of gravity (marked lactometer degrees), and on the slide a scale of temperature ; an arrow is placed at 60° F. and this is placed against the degrees of gravity found ; the temperature then corresponds to the specific gravity corrected to 60° F. ; the readings from the milk scale only deviate from those given by the Table within the limits of reading a lactometer. The dairy scale gives both Fahrenheit and Centigrade degrees.

Collins has devised a milk scale (Fig. 25) by which the percentage of solids not fat can be read direct. In this it is only necessary to set the reading on the lactometer opposite the temperature at which the specific gravity was taken, when the percentage

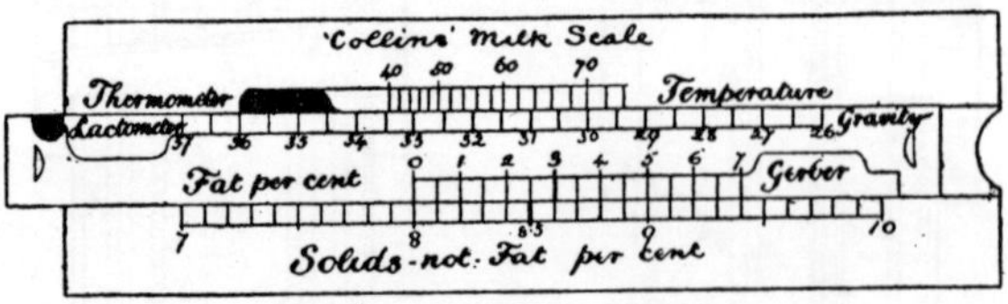

Fig. 25.—Collins' Milk Scale.

of solids not fat will be found opposite the percentage of fat.

Though the calculation of the solids not fat or fat can never be so exact as the direct estimation, it is sufficient for many purposes, where small deviations do not affect the conclusions drawn from the results ; this method also provides a useful check on the analyses where all three estimations are made, and it is hardly ever found that the deviations of the calculated figures from those estimated exceed 0·2 per cent.

Freezing Point.—The apparatus consists of a test-tube $\frac{3}{4}$ inch wide by 6 inches long fitted inside another tube slightly over an inch wide and 5 inches long by means of a cork ring, the outer tube being immersed in a stout glass beaker in which a mixture of 3 parts crushed ice and 1 part of salt is placed; the temperature of the ice and salt mixture should be not lower than $-4°$ C.; a thermometer reading to $\frac{1}{100}°$ C. is suspended in the tube, clamped to a stand, and furnished with a glass rod with a thick india-rubber tube round it, by means of which it can be tapped; a stirrer is also provided, which may conveniently be suspended from the clamp by means of a flexible wire and a small india-rubber ring, and the end of the tapping rod may also be connected with this, so that the same movement works the stirrer and taps the thermometer. A small test-tube containing a drop or two of water, and a thin glass rod is also immersed in the freezing mixture. 20 c.c. of milk are placed in the inner test-tube, and the stirrer worked at a uniform rate of speed till the temperature has fallen to $-1°$ C., when the glass rod coated with a little ice is just plunged into the milk, and removed. The temperature will at once rise, and the highest point is noted, the uniform stirring being continued. The freezing point of a solution of 9·50 grammes of cane-sugar in 100 grammes of water, of which 20 c.c. are also taken, is determined in precisely the same manner, the temperature of the freezing mixture, the point at which the solution is touched off, and the rate of stirring being exactly the same; the difference between the observed freezing point of the cane sugar solution, and $-0{\cdot}535°$ gives the correction to be made to the freezing point of the milk, the numerous corrections otherwise requisite thus being automatically made. The method requires some little skill and practice.

CHAPTER III.

THE ANALYSIS OF MILK-PRODUCTS.

THE liquid milk-products are skim-milk, cream, whey, butter-milk, sterilised milk, condensed milk, and sour or fermented milk ; milk powder consists of milk evaporated to dryness.

Skim-milk is treated exactly as milk ; as the fat globules are very small and few, the estimation of fat requires more care ; the period of revolution of the disc in the Gerber method should be increased, and a correction of 0·05 per cent. should be added to the reading unless a special tube is used. The gravimetric methods of Storch and Ritthausen tend to give slightly low results, and those of Werner-Schmid or Gottlieb are preferable. In the estimation of the proteins by Ritthausen's method, the extraction of the fat may be omitted, and the percentage of fat found subtracted from the results.

Cream requires several modifications. The *specific gravity*, except of a thin cream, is difficult to estimate, and this is usually omitted. It is advantageous to add an equal volume of alcohol to the cream before drying for the *total solid estimation* as there is then no skin to be broken. The *fat* may be estimated by macerating the total solids with ether or amyl alcohol, carefully decanting and repeating the maceration, etc., about half a dozen times ; the solids not fat are weighed directly, and the fat found by difference. The *ash estimation* is made on the solids not fat. The Werner-Schmid method is, however, excellent for cream, with the modification that 2 or 3 grammes should be weighed, into the Stokes tube, and the weight made up to 10 grammes with water. The Gottlieb method is also good; 1 to 2 grammes should be weighed, and enough water added to make the total weight up to 5·15 grammes. The methods are then carried out as for milk.

The fat may be calculated from the total solids with a close approach to accuracy by the formula

$$F = 1{\cdot}102\,T - 10{\cdot}2,$$

or by the following Table :—

TABLE V.

FOR CALCULATING PERCENTAGE OF FAT IN CREAM FROM TOTAL SOLIDS.

Total Solids	Fat	Solids not Fat	Total Solids	Fat	Solids not Fat
60	55·9	4·1	44	38·3	5·7
59	54·8	4·2	43	37·2	5·8
58	53·7	4·3	42	36·1	5·9
57	52·6	4·4	41	35·0	6·0
56	51·5	4·5	40	33·9	6·1
55	50·4	4·6	39	32·8	6·2
54	49·3	4·7	38	31·7	6·3
53	48·2	4·8	37	30·6	6·4
52	47·1	4·9	36	29·5	6·5
51	46·0	5·0	35	28·4	6·6
50	44·9	5·1	34	27·3	6·7
49	43·8	5·2	33	26·2	6·8
48	42·7	5·3	32	25·1	6·9
47	41·6	5·4	31	24·0	7·0
46	40·5	5·5	30	22·9	7·1
45	39·4	5·6	29	21·8	7·2

The fat and solids not fat in cream can be read off directly from the total solids on the dairy scale.

This method is not available in the case of clotted or Devonshire cream.

The aldehyde figure multiplied by 0·45 will approximate to the solids not fat in both fresh and clotted creams.

The **milk-sugar estimation** by the polariscope requires modification, as the cream must be more highly diluted ; it is best to weigh out 50 grammes of cream, make up to 100 c.c., and add 1 c.c. of Wiley's acid mercuric nitrate, and polarise.

The milk-sugar is found by the formula—

$$\text{Milk sugar} = R \times \frac{1}{1{\cdot}05} \times 0{\cdot}95 \times \frac{100 - 1{\cdot}076\,F \times W.}{W}$$

Where R = Reading in angular degrees with a 200 mm. tube.

F = percentage of fat by weight.

W = weight of cream taken.

The gravimetric estimation will require slight correction for the volume of the fat, and the formula—

$$\text{Milk-sugar} = 2M \times \frac{100 - 1{\cdot}076\,F \times W}{W} \text{ may be used.}$$

M = weight of milk-sugar obtained from the Table on p. 34.

The **protein estimation** is carried out as for milk ; it saves time, however, to dry the cream and to extract the bulk of the fat before submitting the sample to Kjeldahl's method, as fat is attacked but slowly by sulphuric acid and mercury.

Butter-milk is analysed in exactly the same way as milk.

It sometimes happens that when churning, both salt and water find their way into the butter-milk ; when the butter-milk is to be sold, it is important to be able to estimate rapidly both the proportion of water and of salt.

Chlorides can be titrated in milk with $\frac{N}{10}$ silver nitrate solution, using potassium chromate as indicator, but as milk itself contains chlorides a correction is necessary for this. To 10 c.c. add a few drops of potassium chromate solution and run in $\frac{N}{10}$ silver nitrate (*see* Appendix) till a faint red tinge is seen; from the quantity used subtract the aldehyde figure (obtained with strontia) multiplied by 0·171; the remainder multiplied by 0·0585 gives the sodium chloride.

By multiplying the amount of salt found by 0·00735 the increment of specific gravity due to the addition is deduced, and subtracting this from the specific gravity found, the specific gravity of the milk is obtained. From this last figure and the fat the solids not fat can be calculated, and from this the amount of added water deduced by the formula on p. 59, or the formula which is based on the above factors,

added water =

$$\frac{36 - \left(G + F - \frac{\text{c.c.}\ \frac{N}{10}\ \text{silver solution} - \text{aldehyde fig.} \times 0{\cdot}171}{}\right.}{}$$

may be used.

Whey is treated as milk; it contains, however, no casein, but gives a small precipitate consisting of albumoses, which by the methods given would be estimated as casein. The aldehyde figure multiplied by 0·125 gives the percentage of proteins.

Sterilised milk can be analysed by the methods given for milk; the polarimetric estimation of milk-sugar tends to be low, owing to change in the milk-sugar on heating, and the gravimetric or volumetric method should be used. The albumin behaves like casein, as it is rendered insoluble in dilute acetic acid and magnesium sulphate solutions; the estimation of casein and albumin can, however, be made by the indirect method from organic phosphorus and sulphur. The total nitrogen is unaffected. Ritthausen's method should not be used for the estimation of fat in sterilised or condensed milk.

Condensed milk, if unsweetened, may be analysed by diluting one part by weight with two parts of water, and boiling and treating in the same way as sterilised milk; the results must, of course, be multiplied by 3.

Sweetened condensed milk should be similarly diluted and analysed. Great care must be taken to mix the contents of the tin well, as the milk-sugar separates in minute crystals, which on long standing sink to the bottom. It is not generally heated to an extent sufficient to affect the milk-sugar, and milk and cane-sugar may be estimated by Harrison's method.

Harrison's Method.—Prepare the filtrate with acid mercuric nitrate as usual (p. 31), taking, however, double quantities, and make the direct polarimetric reading without delay; place as much of the filtrate as possible in a 100 c.c. flask and weigh this; immerse for exactly seven minutes in a briskly boiling water-bath to invert the cane-sugar, cool, and make up with distilled water to the original weight. Fill the 200 c.c. tube with this solution (cleared if necessary by filtration), polarise and at once note the temperature (t). Multiply the difference between the direct and the inverted readings by $\frac{100}{142{\cdot}66 - t/2}$, and this will give the reading due to cane-sugar; this divided by 1·2 will give the percentage of cane-sugar in the diluted milk; the percentage of anhydrous milk-sugar in the diluted milk is given by the difference between the direct reading and that due to cane-sugar. These percentages multiplied by 3 will give the amounts in the condensed milk.

The Werner-Schmid and Ritthausen methods for the estimation of fat should not be used, and the Gottlieb method is recommended.

Sour milk is difficult to analyse, and the results are generally less satisfactory than those obtained with fresh milk. If approximate results only are wanted, such as would be furnished by a determination of specific gravity and fat alone, the following modification of Weibull's method may be used; measure the sour milk, and to each 100 c.c. add 5 c.c. of a solution of ammonia (1 part of ammonia, sp. gr. 0·880, to 4 parts water); shake gently, and allow to stand till the precipitated casein is all dissolved; the specific gravity is estimated by a lactometer, and a correction (usually about 2·7 to 2·8 degrees), found experimentally by noting the decrease of specific gravity in fresh milk treated similarly, is added; the fat is estimated by the Gerber method (or one of the gravimetric methods), and the result is increased by one-twentieth; this method is available if the milk be not too old, and serves excellently for control work.

An estimation of total solids may be made after neutralising the acid with strontia solution; an estimation of the acidity is made as usual (p. 15), a weighed quantity being used instead of a measured amount, and 2 c.c. per 10 grammes less than a proportionate amount of strontia solution added to the weighed quantity of milk taken for total solid estimation; from the weight of total solids 0·00428 gramme should be deducted for each cubic centimetre of $\frac{N}{10}$ strontia added.

The ash is estimated as usual, but 0·00738 gramme should be deducted from the weight of the ash for each cubic centimetre of $\frac{N}{10}$ strontia solution added.

The fat estimation is preferably made by the Storch method (an addition of strontia solution being made as in the total solid estimation) or the Gottlieb method; the Werner-Schmid method is also available, though it tends to give high results with very sour samples owing to the solubility of lactic acid in ether. If, however, the ether be diluted with an equal bulk of petroleum ether in a separating funnel, 10 c.c. of water added and a drop of phenolphthalein solution and then dilute caustic soda solution drop by drop with constant shaking till the aqueous layer be just tinged pink, the lactic acid and other impurities can be removed. The aqueous layer should be rejected, and the ethereal layer washed twice with water, evaporated and the fat weighed.

Other determinations are made as for milk, except that the quantities taken are all weighed and not measured. The total nitrogen is a useful datum.

The aldehyde method does not give exact results for proteins.

The methods used in the Government laboratory include the determination of alcohol, volatile acids, and ammonia, and from these, the solids lost by the various fermentations undergone by the milk are reconstructed. The following description is condensed from Dr. Thorpe's report :—

Alcohol.—To 75 grammes of sour milk half the caustic soda solution necessary to neutralise is added, and the mixture is distilled; to the distillate is added 0·5 c.c. $\frac{N}{10}$ soda solution, and this mixture again distilled; the final distillate is made up to the original bulk, and the specific gravity estimated. The difference in degrees of gravity between the specific gravity and 1,000 multiplied by 0·977 gives the percentage by weight of milk-sugar converted into alcohol.

Volatile Acid.—Ten grammes of milk are neutralised to the extent of one-half, and a little phenolphthalein added; the mixture is evaporated to dryness on a water-bath with frequent stirring, and after the addition of 20 c.c. boiling distilled water, $\frac{N}{10}$ soda solution is added till a pink colour just appears. The difference between the number of c.c. of $\frac{N}{10}$ soda solution used in this experiment, and that required for the original acidity of 10 grammes of milk, is multiplied by 0·0255 to give the percentage of milk-sugar converted into volatile acid.

Ammonia.—Two grammes of milk are diluted to 100 c.c. and filtered clear. Ten c.c. of the filtrate made up to 50 c.c. with distilled water are compared in tint with a solution of ammonium chloride solution (1 c.c. = 0·01 milligramme NH_3) in 50 c.c. water containing 10 c.c. of a solution of 2 grammes fresh milk acidified in 100 c.c. after the addition of 2 c.c. of Nessler solution (*see* Appendix) to each. The number of c.c. of ammonia solution required to produce the same tint multiplied by 0·026 gives the percentage of casein converted into ammonia.

The three amounts are added together, and constitute the total correction for solids lost by fermentation.

Miller and the author have suggested other corrections, which are described at length in *Dairy Chemistry*, but in most cases the above will suffice.

Milk Powder.—Moisture is estimated by drying

1 to 2 grammes in a basin to constant weight, and the ash is estimated in the same portion.

The estimation of fat should be made by the Gottlieb method, 0·6 to 0·7 gramme being weighed out and water sufficient to make up to 5·15 grammes; after the addition of the alcohol, the solution may be warmed if necessary to effect complete solution, and cooled before the ether is added.

For the estimation of nitrogen by Kjeldahl's method about 1 gramme should be taken.

For other estimations 10 grammes may be dissolved in water, and made up to 100 c.c. after heating and cooling.

CHAPTER IV.

THE APPLICATION OF ANALYSIS TO THE SOLUTION OF PROBLEMS.

The Detection of Adulteration.—The principal forms of adulteration of milk are the addition of water and the removal of cream.

The detection of water is based on the reasoning that while the water natural to milk contains solids not fat, added water is free from these. The amount of solids not fat is nearly though not quite constant, and rarely falls below 8·5 per cent. or rises much above 9·2 per cent.; numerous cases, however, are on record of solids not fat below 8·5.

The removal of cream is detected by a deficiency in the fat; this varies much more than the solids not fat, but comparatively rarely falls below 3·0 per cent.

TABLE VI.

Percentage of Solids not Fat	Number of Samples	Percentage of Fat	Number of Samples
8·4 to 8·5	1892	2·9 to 3·0	370
8·3 to 8·4	242	2·8 to 2·9	209
8·2 to 8·3	27	2·7 to 2·8	87
8·1 to 8·2	22	2·6 to 2·7	37
8·0 to 8·1	8	2·5 to 2·6	16
Below 8·0	2	Below 2·5	13

The probability of samples falling below 8·5 per cent. of solids not fat, and 3·0 per cent. of fat is indicated by the foregoing Table, which gives the number of samples per 100,000 which may be expected at each percentage named; it is assumed that each sample represents a churn of milk—*i.e.*, that the milk is the mixed product of several cows.

By Clause 4 of the Sale of Food and Drugs Act, 1899, the President of the Board of Agriculture is empowered to lay down limits below which a presumption is raised that milk is not genuine, and he has fixed 8·5 per cent. of solids not fat and 3·0 per cent. of fat. The effect of this is that the onus of proving that milk taken under the Sale of Food and Drugs Acts falling below these limits is genuine lies on the vendor, and for most practical purposes milk below these limits is taken as adulterated.

It is generally, though not invariably, found that, in milk falling below 8·5 per cent. of solids not fat, the deficiency lies chiefly on the milk-sugar, and that the proteins and ash are normal; a percentage of total nitrogen above 0·5 and a percentage of ash above 0·7 in a milk below 8·5 per cent. of solids not fat will afford strong evidence that the milk is genuine, while figures for total nitrogen and ash low proportionately to the solids not fat will strengthen the conclusion that the milk is watered. Abnormal milks low in solids not fat have been found to have a normal freezing point—*i.e.*, not less than − 0·508° C.

There appears to be no chemical means of distinguishing between fat naturally low and fat lowered by the abstraction of cream; the most numerous instances of fat below 3·0 per cent. naturally occurring have been found in April, May, June, and July, and they are especially rare in October, November, and December.

The percentage of added water may be calculated by the formula :—

$$\text{Added water} = \frac{8{\cdot}5 - S}{8{\cdot}5} \times 100. \quad (S = \text{solids not fat.})$$

A formula which gives a nearer approach to the probable amount is :—

$$\text{Added water} = \frac{36 - (G + F)}{36} \times 100.$$

(G = degrees of gravity. F = fat.)

The percentage of added water may be calculated from the aldehyde figure (A) by the formula :—

$$\text{Added water} = \frac{20 - A}{20} \times 100.$$

The amount of cream abstracted may be calculated by the formula :—

$$\text{Cream abstracted} = \frac{3 - F}{3} \times 100. \quad (F = \text{fat.})$$

This gives the minimum percentage of cream abstracted, and the more probable amount is obtained by substituting the average percentage for the month as given in Chap. i. (p. 4) for 3 in both places.

The above calculations may be made by the dairy scale.

Detection of Preservatives—*Boric Acid.*—To a little milk add a few drops of phenolphthalein, and caustic soda solution drop by drop till a faint pink colour is produced ; place some of the milk in two test-tubes, dilute one with an equal volume of water, and the other with a *neutral* 50 per cent. solution of glycerin ; in the absence of boric acid the two tubes will have almost the same colour, in its presence the glycerin tube will be lighter, and usually white.

As an alternative method the milk or its ash

may be made distinctly, but not strongly, acid with hydrochloric acid and a piece of turmeric paper dipped into the solution: on drying, the paper turns pink in the presence of boric acid, and is turned a greenish-black by alkalies.

ESTIMATION OF BORIC ACID.—To 10 c.c. of milk add at least 5 c.c. of phenolphthalein solution; raise to the boil, and neutralise with $\frac{N}{10}$ alkali while still boiling: the pink colour produced when neutral is faint, though quite distinct. Add 8 to 10 c.c. of glycerin, and run in the alkali till a pink colour appears, boiling at this stage being unnecessary; the quantity of $\frac{N}{10}$ alkali used, corrected for the acidity of the glycerin, multiplied by 0·062 gives the percentage of boric acid (calculated as H_3BO_3).

Formaldehyde.—Dilute a little milk with an equal bulk of water in a test-tube; pour carefully down the side of the tube a little 90 per cent. commercial sulphuric acid; a bluish colour is developed at the junction of the acid and milk in the presence of formaldehyde. This blue colour may also be observed during the estimation of fat by the Gerber process (p. 16).

Salicylic and Benzoic Acids.—These acids are best tested for by making the milk (or cream) alkaline with sodium carbonate, adding one-tenth of the volume of 10 per cent. calcium chloride solution, and heating on the water-bath to separate the casein; after cooling the filtrate is neutralised, and the proteins removed as in Ritthausen method (p. 35). The filtrate from this is acidified, and extracted with a mixture of ether and petroleum ether, which should be separated, washed, and shaken with a little water to which a drop of phenolphthalein solution has been added, dropping in dilute caustic soda solution, till the water after shaking just turns pink ; after discharging the pink colour with very dilute acetic acid and adding a little ferric chloride solution, the presence of salicylic acid is shown by a violet colour and of benzoic acid by a buff precipitate.

Hydrogen Peroxide.—Mix the sample with a little fresh milk, and add a small amount of para-phenylenediamine or ortol ; a blue or red colour will be developed if hydrogen peroxide be present.

Reaction of Milk with Hydrogen Peroxide.—Fresh milk when treated with a little para-phenylenediamine or ortol (a photographic developer) and a drop of hydrogen peroxide gives a deep blue (with the diamine) or a brick-red (with ortol) coloration within a few seconds. Milk heated above 80° C. remains white.

The Cause of Poor Milk.—The detection of added water and of a deficiency of cream would be an obvious explanation of the cause of milk being poor. If it be normal in composition, but very white, and the fat separated in the Gerber process is nearly free from colour, this would show that the poverty of the milk had been inferred fallaciously from its lack of colour ; if this be not the case, the test given above should be applied, and the soluble albumin estimated (p. 36) ; a deficiency of albumin below 0·35 per cent., or the non-production of colour with para-phenylenediamine or ortol will show that the milk has been heated, and as the cream rises very slowly on heated milk, the milk has been called poor because cream is not apparent in a short time. Occasionally a sample called poor turns out to contain a high percentage of fat ; this would show that the milk has been standing long enough for the cream to separate and it is then divided into a rich and a poor portion.

A not unusual practice is for the servants in a household to pour off a portion of milk from a can which has stood some time for their own consumption, and remove the cream, and to send the rest of the milk, thus impoverished, for the consumption of the other members of the household.

The Cause of Sweet Milk.—Milk is sometimes alleged to be sweet. Cane-sugar may be detected by adding to 10 c.c. of milk 0·1 gramme of resorcinol and 1 c.c. of strong hydrochloric acid; on standing in boiling water for five minutes a red colour is produced if cane-sugar be present; it may be estimated by Harrison's method (p. 52). If all the figures for solids not fat, ash, sugar, and proteins be equally high, the milk has simply been concentrated, usually by boiling; the solids not fat have been found as high as 17·5 per cent. in a case of this kind.

The Cause of High Colour.—If the colour be yellow and the fat separated by the Gerber process very much darker than usual, and the cream separating be much yellower than the skim-milk, the high colour is natural.

If this be not the case artificial colours should be tested for; annatto is detected by making the milk alkaline with sodium bicarbonate, immersing a strip of filter-paper in it, and allowing to stand till next day; in the presence of annatto the strip is stained brownish. Coal-tar dyes of the azo group give a pink colour when a mineral acid is added to milk, and this is usually seen in the Gerber test. Other artificial colours are practically never used.

A pink colour is generally due to blood; to detect this the milk is warmed to 50° C., and separated in a high-speed centrifuge (Fig. 26); a bright red deposit at the bottom of the tube may be taken

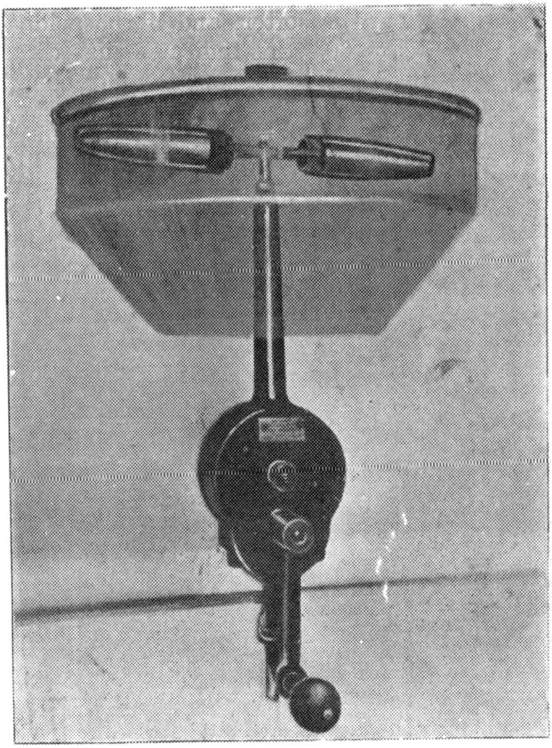

Fig. 26.—High-speed Centrifuge.

as blood; the microscopic appearance of the corpuscles is given in Fig. 27.

The Cause of Sour Milk.—Practically the only cause of milk turning sour is the formation of lactic acid by the action of micro-organisms. Milk curdles on boiling when the acidity reaches about 33°, and spontaneously in the cold when the acidity is about

80° ; milk curdled by boiling generally contains somewhat hard lumps of curd, and the acidity of the whey may be 25° and upwards ; at lower temperatures the curd is softer, and if the acidity be appreciably below 80°, this indicates that the milk has been kept warm, but not warm enough to inhibit microbial action.

Freshness of Milk.—The freshness of milk may be determined by an acidity estimation ; if the acidity be more than 2° higher than the aldehyde

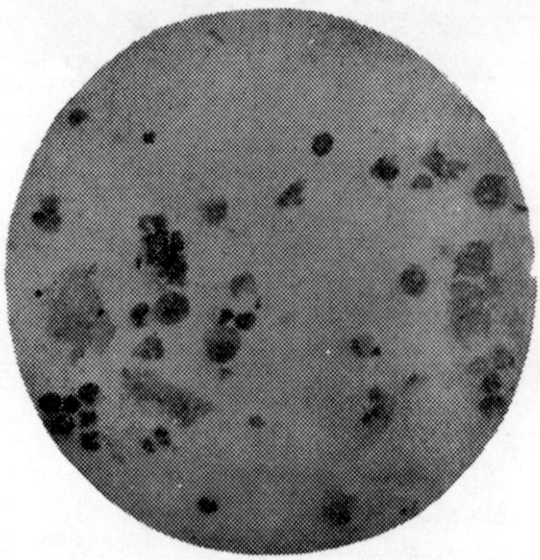

Fig. 27.—Blood in Milk.

figure, it may be assumed that a development of acidity due to the action of micro-organisms has taken place.

Many micro-organisms secrete a reductase, which decolourises methylene blue ; make a solution by diluting 1 c.c. of saturated alcoholic solution of methylene blue to 100 c.c. with distilled water, add 1 c.c. of this to 25 c.c. of milk contained in a stoppered tube of little more than 25 c.c. capacity, and keep for half an hour at a temperature of 37° C.

(blood heat); if the colour disappears, the milk is not fresh.

Milk which is not fresh gives a high catalase figure, but this is not conclusive evidence, as high catalase figures are obtained with fresh milk from cows with diseased udders.

Custards made with eggs, especially if much sugar be added, inevitably curdle if heated to too high a temperature; the liquids from such samples are usually high in total solids (about 16 per cent.), have a low acidity, and are nearly clear, characteristics which permit of the cause of curdling being established.

Occasionally milk is alleged to be sour because it turns blue litmus red; all milk does this, as well as turning red litmus blue, owing to the amphoteric reaction of the phosphates and citrates of the milk.

The three popular superstitions that milk can be kept warm—*i.e.*, a little above blood heat for several hours without change, that custards can be heated to any temperature, and that blue litmus can be used as a test for sour milk, though all fallacious, are responsible for many allegations of sour milk. The peptonisation of milk with powders which are insufficiently alkaline or which have been kept too long may cause it to curdle, and to be thought to be sour.

The Cause of Unusual Taste.—Milk on boiling acquires a taste, and the tests for heated milk (p. 61) and soluble albumin (p. 36) will show this. Mixture with dirty water may give an evil taste to milk, and usually the establishment of the presence of water is all that can be done to explain this; if an alkali be added to milk, the taste is soapy and the smell fishy, and an increase in ash and its strongly marked alkalinity will detect the cause. If the taste be due to a fermentation other than the

5

normal lactic one, or to the food of the cattle, it is usually difficult to detect the cause by chemical analysis.

An unpleasant taste may be due to the presence of urine, accidentally or wilfully added.

Fig. 28.—Using Microscope.

Detection of Urine.—Half fill a small test-tube ($2 \times \frac{1}{4}$ is large enough) with sodium hypobromite solution (*see* Appendix); carefully fill the tube with milk so that the two liquids do not mix. Place the thumb over the tube, and invert once or twice, and then hold it, with the thumb still over the opening, upside down. Milk causes practically no pressure on the thumb, and gives not more than one-fifth of its volume of gas ; if urine has been added, much pressure is developed, and the liquid squirts out ; milk containing 1 per cent. of urine yields about two-fifths the volume of gas, while 5 per cent. causes an evolution of gas equal in volume to the milk taken. Lobeck's catalase tubes (p. 41) may be used.

The Cause of Dirty Milk. Dirty milk almost invariably deposits a sediment on standing or centrifuging ; the milk is carefully decanted, the sediment washed with water, and allowed to settle again and examined under the microscope (Fig. 28). Sharply defined vegetable cells (Fig. 29), indicate that the finer particles of the food given to the cattle probably at milking time have fallen in ; less well defined vegetable cells (Fig. 30), stained yellowish are probably derived from fæcal matter ; small hairs and various fibres (cotton, wool, etc.) show the presence of household dust ; transparent irregular particles which do not polarise are quartz, and are due to road dust ; this latter also gives a strong reaction for iron on treating the sediment with hydrochloric acid, diluting, and testing with potassium ferrocyanide which gives a blue colour with iron salts.

Control of Milk Prescriptions.—To control the preparation of milk mixtures made up for infant feeding from prescriptions, it is necessary to estimate rapidly fat, proteins, milk-sugar, and added sugar. The problem is often simplified by all the materials used being available for analysis, and by determining fat (by the Gerber method), specific gravity, and aldehyde figure, and calculating the

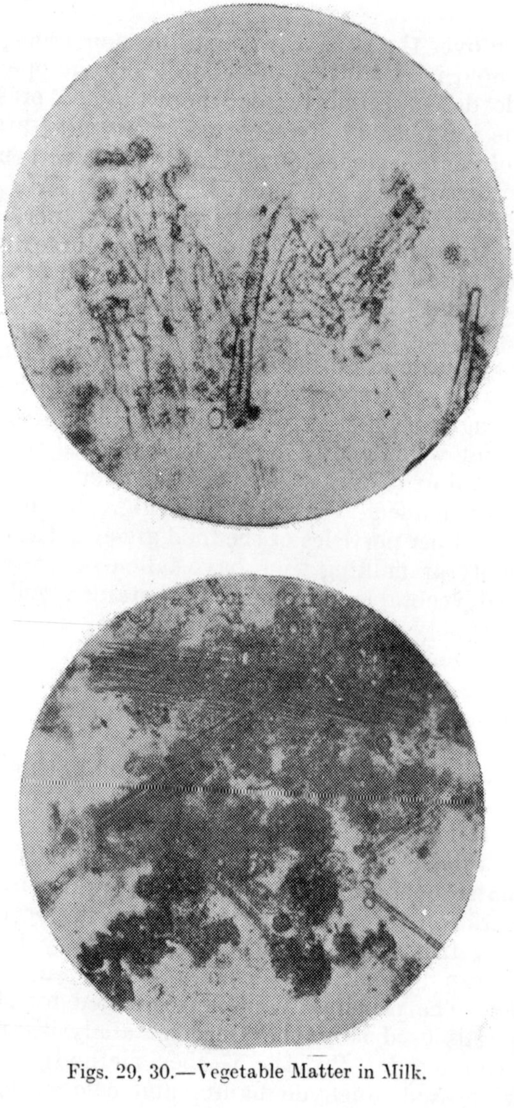

Figs. 29, 30.—Vegetable Matter in Milk.

solids not fat, in the samples and in the milk from which the samples were made, the four determinations required can be obtained with sufficient accuracy for control purposes.

By estimating the ratio of solids not fat to aldehyde figure in the milk used, and multiplying the aldehyde figure of the mixture by this ratio, the amount of solids not fat derived from the milk is obtained, and the difference between the actual amount and this gives the added sugar. With an unknown milk mixture the ratio 0·45 may be assumed. The method does not, of course, distinguish between added cane-sugar or added milk-sugar, but when it is known what the added substance is, an estimation of sufficient accuracy is obtained.

Detection of Adulteration of Cream.—As there is no standard for cream, it may contain any percentage of fat, and still be cream ; there is a practical standard of " thickness " which the purchaser mentally estimates, and judges the value of the cream thereby. Artificial thickening is sometimes resorted to, and gelatinised starch, gelatin, " viscogen " (a solution of lime in cane-sugar syrup), condensed milk or milk powder, and agar are added.

Starch is detected by the blue colour produced on adding a solution of iodine in potassium iodide.

Gelatine is found, if present, by diluting the cream with water and adding a little acid mercuric nitrate solution (*see* Appendix) ; the filtrate, if gelatine be present, is usually turbid, and gives a precipitate on the addition of a saturated solution of picric acid.

Viscogen raises the percentage of lime in the ash ; the lime on an average amounts to 22 per cent. of the ash, and its ratio to phosphoric acid (CaO to

P_2O_5) is 1 : 1·3. Viscogen raises not only the percentage in the ash, but also the ratio to the solids not fat. A small cane-sugar reaction may be obtained, and the percentage of sugar polarised as milk-sugar will exceed 52·5 per cent. of the solids not fat.

Condensed milk or *milk powder* may be detected by the solids not fat being found in much greater proportion than that given as corresponding to the fat found in Table V.; the percentages of ash, milk-sugar, and proteins, and the aldehyde figure will bear the same proportion to the solids not fat as found in milk. Clotted or Devonshire cream, however, is concentrated during its preparation, but its physical appearance differs from that of raw cream, and it does not give the reactions with hydrogen peroxide given on p. 61.

Agar is difficult of detection; it raises the percentage of solids not fat, and gives the same reaction as cane-sugar with resorcinol (p. 62), but by Harrison's method no cane-sugar is shown.

Preservatives are detected in the same way as for milk. The Cream Regulations allow as percentage of boric acid not exceeding 0·4 per cent to be declared. For the estimation of boric acid, the cream should be diluted with an equal weight of water.

Adulteration of Skim-milk.—The President of the Board of Agriculture has fixed 8·75 per cent. as the limit for solids not fat in skim-milk. Percentages below this are presumed to be caused by the addition of water.

Rennet is sometimes added to skim-milk, and even to whole milk, usually with the idea of causing curdling when the milk is warmed. Its presence may be inferred if the milk curdle on warming to 40° C., and the acidity be less than 25°; the whey on neutralizing to an acidity of 12°, and mixed with an equal volume of milk, causes curdling when warmed to 40° C., and the amount of lime in the whey does not exceed 0·06 per cent.

Detection of Foreign Fats in Milk and Cream.—By means of an emulsifying apparatus, foreign fats (margarine fat, coco-nut oil) are mixed with separated milk and the product sold as milk or cream. The casein should be precipitated from a considerable amount of milk (p. 36), dried, extracted with ether, and the fat examined as butter fat (pp. 75, *et seq.*).

CHAPTER V.

THE ANALYSIS OF BUTTER.

Estimation of Water.—Weigh a small round basin (Fig. 31) about 3 inches in diameter, containing a

Fig. 31.—Water Estimation.

small rod; place 5 to 10 grammes of butter therein, and weigh again; heat the basin over a very small flame, or on a sand-bath, and stir *constantly* till frothing has ceased, cool, and weigh again. The loss of weight indicates water, and this multiplied by 100 and divided by the weight of butter taken gives the percentage. The flame should be of such a size that the butter takes at least a minute to become melted.

As an alternative method, about $2\frac{1}{2}$ grammes of butter may be weighed in a flat-bottomed basin, just melted in the water-oven, and $1\frac{1}{2}$ c.c. strong alcohol mixed with the melted fat; the basin is placed in the water-oven for two hours, cooled, and weighed. The loss of weight gives the water.

Estimation of Curd and Salt.—Wash the fat from the basin after driving off the water by nearly filling it with ether or amyl alcohol, and carefully decanting the liquid after the solid portion has settled, and repeating this four times; if amyl alcohol be used, it should be hot; the residue is dried in the water-oven for two hours and weighed after cooling. This represents curd and salt (if present).

Extract the salt from the curd with hot water, and filter the solution; wash the residue and the filter, and cool the filtrate; add a few drops of potassium chromate solution, and titrate with $\frac{N}{10}$ silver nitrate solution till a faint red colour just appears; each cubic centimetre used is equal to 0·00585 gramme salt.

Estimation of Casein.—Extract another portion of curd with dilute ammonia till no lumps are left; filter and wash the residue; add dilute acetic acid till a white precipitate falls, and collect this in a weighed Gooch crucible or on tared filter-papers, as on p. 36. Extraction and ignition may be omitted. The precipitate is casein.

Tests for Preservatives.—Boric acid may be detected by melting the butter at a low temperature, and testing the aqueous portion with turmeric paper as directed on p. 59.

If found, the estimation is carried out by weighing 25 grammes of butter, adding 25 c.c. of water, melting the butter at a low temperature, and stirring well; the aqueous portion is allowed to settle, and 20 c.c. are withdrawn, and placed in a small beaker and boiled; at least 10 c.c. of phenolphthalein solution (*see* Appendix) is added, and the solution titrated while still boiling with $\frac{N}{2}$ caustic soda solution (*see* Appendix) till a faint pink colour occurs; the reading of the burette is noted, 12 c.c. glycerin are added, and the solution titrated again (boiling is not necessary) till a pink colour appears; the difference between the reading of the burette and the first reading, corrected for the acidity of the glycerin, multiplied by 0·031, will give the weight of boric acid (as H_3BO_3), and this multiplied by $5 + 0{\cdot}05$ W (W = percentage of water) will give the percentage of boric acid. Where the percentage of water is not estimated the figure 5·65 may be used for $5 + 0{\cdot}05$ W, and as an approximate figure the number of cubic centimetres of $\frac{N}{2}$ soda solution used between the first and second titrations may be multiplied by 0·17.

Sulphites are detected by the smell of sulphurous acid given off on acidifying the aqueous portion; formaldehyde will give its characteristic reaction (p. 60) if a little of the aqueous portion be added to milk; fluorides are detected by shaking 10 grammes after melting with ether and 1 or 2 c.c. of water in a separating funnel. The aqueous layer is run off into a test-tube, a few drops of hydrogen peroxide added, and 1 c.c. of a solution containing 2 per cent. of titanium sulphate in 10 per cent. sulphuric acid. In the presence of fluorides the orange-yellow colour will be partially or wholly discharged. A blank test on pure butter should be made for comparison.

The Examination of the Fat.—Butter fat is of peculiar composition, consisting of complex glycerides containing lower fatty acids, chiefly caproic and butyric acids; these are characterised by being soluble in water, and volatile with steam, while the fatty acids of almost all other fats are insoluble and non-volatile; furthermore the presence of the lower fatty acids in the glycerides causes them to have a softer consistency than if only the insoluble acids were present, and a comparatively small amount of the acids of the oleic and more unsaturated series is present. To obtain a fat of the consistency of butter without the lower fatty acids, a larger amount of acids of the oleic, etc., series must be present. The presence of the lower fatty acids gives a high specific gravity to the glycerides, and causes them to crystallise badly.

In addition to these facts on which the broad principles of butter analysis are based, certain vegetable oils, especially sesamé oil, give characteristic reactions, and it has been recommended that by international agreement all margarine shall legally be made to contain 10 per cent. of sesamé oil.

The addition of margarine to butter may be detected by—

(*a*) A lowered proportion of volatile acids;
(*b*) A lowered proportion of soluble and increased proportion of insoluble acids;
(*c*) An increased mean molecular weight;
(*d*) A decreased density;
(*e*) A more marked crystallisation;

properties all chiefly depending on the lowering of

the amount of caproic and butyric acids in the glycerides, and—

(*f*) An increased iodine absorption;
(*g*) An increased refractive index;

properties chiefly depending on the increase of the unsaturated acids in the glycerides; to those may be added—

(*h*) A turbidity of the fat on melting at a low temperature, a property which depends on the fact that mixing margarine with butter often causes overworking, which gives rise to turbidity.

Owing the the natural variations of the composition of butter fat, and to a less degree of the composition of adulterants, the detection of small quantities of margarine is difficult, and unless sesamé oil can be found, impossible in minimal amounts; as only three principles underlie all the methods, one of which has little value, a multiplication of tests does not greatly assist.

Preparation of the Fat for Analysis.—Place 20 to 50 grammes of butter in a small beaker, and put this in the water-oven till melted; observe the fatty layer, whether clear or turbid; pour as much as possible of the fat into a dry filter, taking care that none of the aqueous portion accompanies it, and filter in the water-oven, collecting the clear fat in a small beaker.

Estimation of the Volatile Fatty Acids—*Reichert-Wollny Process, Polenske's Modification.*—Place a flask of 300 c.c. capacity on one pan of a balance and tare it; add 4·5 grammes to the weights, and run in the melted fat till the flask is weighed down, and place a further 0·5 gramme weight on the other pan; continue the addition of the fat cautiously, till the weight is exact, if necessary removing a surplus with a small pipette. It is not necessary to wait for the fat in the flask to cool before making the final adjustment, as the error involved in weighing warm fat is within the limits of error of the final titration, nor it is necessary to weigh more accu-

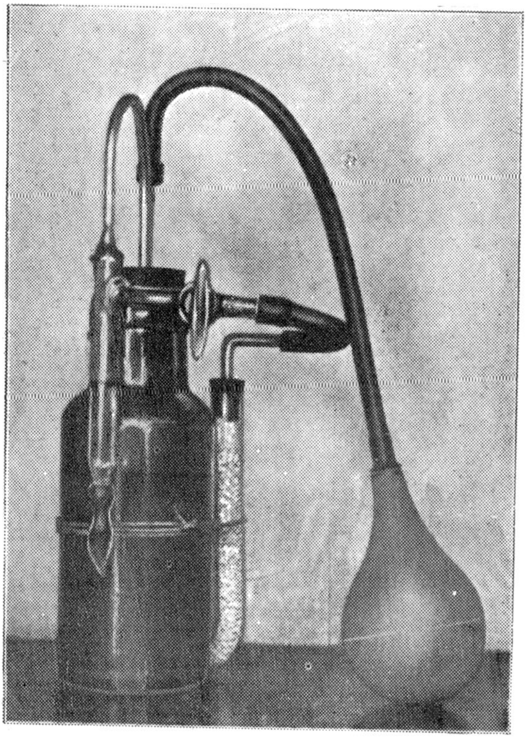

Fig. 32.—Caustic Soda Apparatus.

rately than 0·005 gramme. Weigh in 20 grammes of glycerin.

Add 2 c.c. of a 50 per cent. solution of caustic soda (*see* Appendix), preferably from a special measuring apparatus (Fig. 32); heat the mixture over a naked flame till the turbid liquid suddenly becomes clear; allow it to cool a little and add 100 c.c. of hot water which has been previously

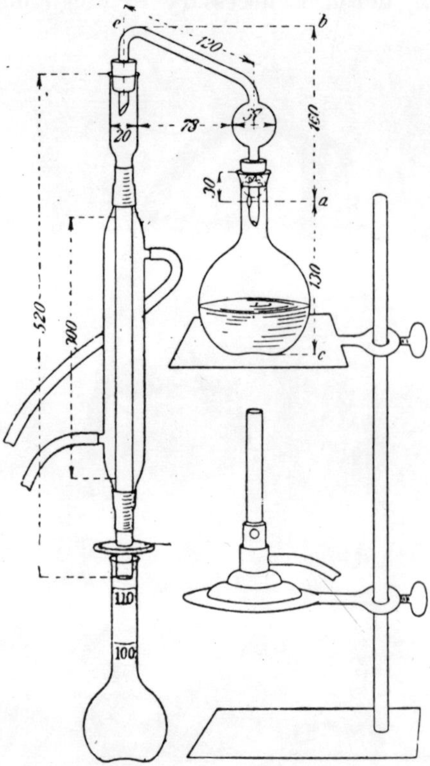

Fig. 33.—Polenske Apparatus.

boiled for at least fifteen minutes, and when all the resulting soap is dissolved 0·1 gramme of powdered pumice which has been sifted through muslin and ignited, and 40 c.c. of sulphuric acid solution (*see*

Appendix); attach the flask by means of a cork to a bulb tube attached to a condenser, the dimensions of which are given in the sketch (Fig. 33). Support the flask on a piece of asbestos card in which is cut a hole 5 c.m. in diameter and heat with a very small flame, till the fatty acids float in a clear layer on the surface of the liquid; then turn up the flame to such a height that 110 c.c. distil in from eighteen to twenty-two minutes. Collect 110 c.c., turn out the flame, and replace the flask by a 25 c.c. cylinder; cool the flask for ten minutes in water at 10° C., mix the distillate, and filter through a dry filter; use the first few c.c. to wash out a 100 c.c. flask, and collect exactly 100 c.c. of the filtrate, and transfer this to a beaker, add a little phenolphthalein and titrate with $\frac{N}{10}$ baryta, strontia, or soda solution till a pink colour appears; pour this back into the flask, and again into the beaker, and if the colour be discharged continue the titration. From the number of cubic centimetres used, subtract the figure obtained in a blank experiment, and multiply the result by 1·1 to obtain the Reichert-Wollny figure.

Wash out the condenser with two successive quantities of 9 c.c. each of cold water, collecting this in the cylinder and using it to wash out the flask, and pouring it through the filter; reject this filtrate, and remove the funnel containing the filtrate to a clean flask. Wash out the insoluble fatty acids from the condenser with three successive quantities of 10 c.c. each of neutral alcohol, collecting each in the cylinder, and pouring it through the filter; titrate the combined filtrates with $\frac{N}{10}$ alkali after adding a little phenolphthalein; the number of c.c. of $\frac{N}{10}$ alkali used corrected for a blank experiment gives the Polenske figure.

The Polenske figure varies with the Reichert-Wollny figure, and the following Table shows the relation :—

TABLE VII.

Reichert-Wollny Figure	Polenske Figure	
	Mean	Maximum
32	3·2	3·7
31	3·0	3·5
30	2·8	3·3
29	2·7	3·2
28	2·6	3·1
27	2·4	2·9
26	2·3	2·8
25	2·1	2·6
24	2·0	2·5
23	1·9	2·4
22	1·8	2·3
21	1·7	2·2

This relation is also given on the dairy scale.

Kirschner Process.—To the 100 c.c. of the distillate titrated to obtain the Reichert-Wollny figure, add 0·5 gramme of finely powdered silver sulphate, and after standing for an hour with occasional shaking filter the liquid. 100 c.c. of the filtrate is placed in the distilling flask, 35 c.c. of well-boiled water added, and 10 c.c. of the sulphuric acid. A long piece of aluminium wire is placed in the flask, and 110 c.c. distilled as before; 100 c.c. of the distillate are titrated with $\frac{N}{10}$ alkali, and after correction for the blank, the Kirschner figure is calculated by multiplying the c.c. of alkali used by 1·21 and by $\frac{100 + x}{100}$, where x = the number of c.c. of alkali added to the first distillate for neutralisation.

The average relation between Polenske and Kirschner figures is given by the formula

$$P = (K - 14) \times 0{\cdot}26,$$

and the maximum never exceeds

$$P = (K - 10) \times 0{\cdot}26.$$

The average relation is shown on the dairy scale.

Estimation of Soluble and Insoluble Fatty Acids and Mean Molecular Weight.—Weigh a glass flask, which has been previously well boiled with caustic alkali solution; add about 4 grammes of butter fat, and weigh again. Run in 50 c.c. $\frac{N}{2}$ alcoholic soda solution (*see* Appendix), attach the flask to an upright condenser, and boil for a quarter of an hour; add a few drops of phenolphthalein solution and titrate with $\frac{N}{2}$ hydrochloric acid till the pink colour is discharged. As the alcoholic soda solution alters in strength it must be checked against the $\frac{N}{2}$ hydrochloric acid, and the value in terms of the acid solution obtained of 50 c.c. of alcoholic soda; from this subtract the volume of $\frac{N}{2}$ hydrochloric acid used in titrating after saponification, and multiply by 28·05 and divide by the weight of butter taken; the figure obtained will be the milligrammes of potash required for saponification of 1 gramme.

Wash out the alcoholic solution into a large basin, and evaporate the alcohol on the water-bath; add enough hot water to make the bulk up to 150 to 200 c.c. and then add sufficient $\frac{N}{2}$ hydrochloric acid solution to make with the volume already used in the titration 1 c.c. more than the quantity equal to the 50 c.c. alcoholic soda added; heat on the water-bath till the insoluble fatty acids float on the surface in a clear layer, and filter through a wet filter; wash out the basin with hot water till the fatty acids are transferred to the filter, and wash them well on the filter, stirring them up with the jet of water; at least a litre of water is required for washing. With a good filter the fatty acids do not run through; when washed, allow all the water to run out, and wash the filter with hot alcohol, collecting the filtrate in a weighed flask, till all the fatty acids are removed; evaporate the alcohol, and dry in the water-oven till the weight is constant; this will give the insoluble fatty acids.

To estimate the soluble fatty acids, add a little phenolphthalein solution to the filtrate from the insoluble fatty acids, and titrate with alcoholic alkali till a pink colour appears; from the volume used calculate the equivalent of $\frac{N}{2}$ acid, from this subtract the 1 c.c. added in excess of the soda added, multiply by 4·4, and divide by the weight of butter taken; this will give the soluble fatty acids in terms of butyric acid.

Avé-Lallemant Method.—Four grammes of fat are saponified and neutralised as directed above, and the alcohol evaporated; 20 c.c. of water are added, and the saponified fat evaporated to dryness. About 350 c.c. of hot water are added, and the soap solution transferred to a 500 c.c. flask, which is placed on a water-bath. 100 c.c. $\frac{N}{2}$ barium chloride solution are added with constant shaking, and the flask is left for fifteen minutes on the water-bath; it is then cooled, and the solution made up to 500 c.c. and filtered; 200 c.c. of the filtrate is raised nearly to the boiling-point, and dilute sulphuric acid added till no further precipitate is obtained, the precipitate collected on a filter, washed with water, alcohol, and ether, ignited and weighed; the difference between the weight and that of the barium sulphate obtained from 40 c.c. of the barium chloride solution, multiplied by 1,643·5 and divided by the weight of fat taken, gives the insoluble baryta value (b). The saponification value is calculated in terms of barium oxide by multiplying the milligrammes of potash required for saponification (*see above*) by 1·368 (a), and the difference between a and b will give the soluble baryta value (c).

In genuine butters the value of $b - (200 + c)$ is always negative, and varies from $-0·7$ to $-23·8$; coco-nut oil, margarine, and other fats give positive values.

The soluble baryta value of butter usually lies between 50 and 65, but may be occasionally higher; coco-nut oil gives nearly the same figure, but the soluble baryta value for other fats does not exceed 10.

Phytosteryl Acetate Method.—Fifty grammes of the clear melted fat are shaken with 75 c.c. of warm 95 per cent. alcohol; the alcohol is cooled and decanted, and a further quantity of 75 c.c. of alcohol is added, and the process repeated.

The fat is saponified with 1 c.c. of 50 per cent. caustic soda solution, and the bulk of the alcohol evaporated in a basin; 2 grammes of sodium bicarbonate and 2 or 3 grammes of kieselguhr are added, and the mass evaporated to dryness, placed in an extraction thimble, and extracted with petroleum ether in a Soxhlet. Evaporate the solution, add 5 c.c. of $\frac{N}{2}$ alcoholic soda solution, and evaporate to dryness, adding a little sodium bicarbonate before the solvent has completely evaporated. Extract with petroleum ether, and evaporate this, and take up with alcohol; if the solution is dark add a little animal charcoal and filter. Allow the cholesterol or phytosterol to crystallise, and to the crystals add 2 or 3 c.c. of acetic anhydride, and evaporate on the water-bath. Crystallise the acetates several times from alcohol and take the melting-point. Cholesteryl acetate melts at 113·2°-114·6° (corr.) and a higher melting-point shows presence of phytosteryl acetate.

Estimation of Density.—Follow the directions on p. 9, with the following modifications:—Weigh the tube full of water at 37·8° C. (100° F.) instead of at 15·5° C.; dry the tube before filling it with the fat, and take the density at 37·8° C. instead of at 15·5° C.

Examination under Polarised Light.—Place a small portion of the butter (not butter fat) on a microscope slide, and press down a cover-glass thereon; examine with a microscope furnished with a polariscope, using a 1 inch or ½ inch power, focus with the Nicols parallel, and then cross them, shielding the slide from light except that which has passed through the polariser. Genuine butter appears nearly uniformly dark, while crystalline fats show a more or less well-marked lighting in portions of the field.

Old butters, especially those which have been submitted to vibrations, and butters prepared by processes in which the cream is churned soon after heating and cooling may show (Fig. 34) a somewhat crystalline appearance, but generally this is due to margarine (Fig. 35); this test though very rapid may not be reliable.

Estimation of Iodine Absorption.—Weigh about 0·4 to 0·5 gramme of fat in a stoppered bottle, dissolve in 10 c.c. of carbon tetrachloride, and add 20 c.c. of Wijs' iodine solution (*see* Appendix). At the same time mix 10 c.c. of carbon tetrachloride with 20 c.c. of iodine solution, and place both bottles in a dark place for half an hour. Add 15 c.c. of a 10 per cent. solution of potassium iodide solution to each, and about 200 c.c. of water, and titrate with $\frac{N}{10}$ sodium thiosulphate solution (*see*

Fig. 34.—Butter.

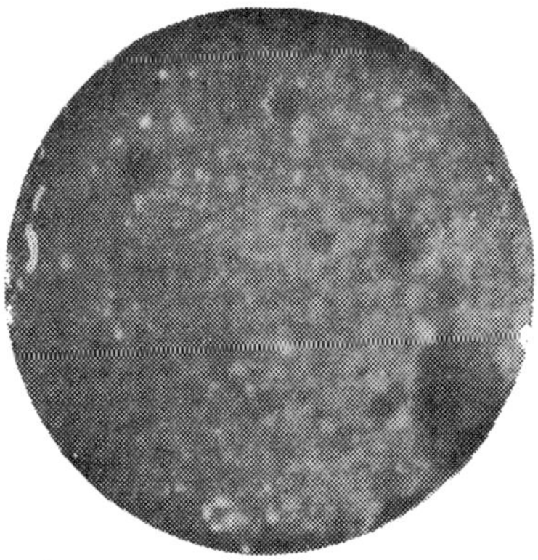

Fig. 35.—Margarine.

Appendix) till the colour on shaking is removed from both aqueous and tetrachloride solutions; a little starch solution may be added when the colour is very pale, and the titration carried on till all blue has disappeared, but its use is not absolutely necessary.

The number of c.c. of thiosulphate multiplied by the value of 1 c.c. in terms of iodine used in the blank experiment gives the total amount of iodine added to the fat (*cf.* p. 107); the number of c.c. of thiosulphate similarly multiplied used in the actual experiment gives the weight of iodine not absorbed by the fat, and the difference between these two gives the quantity absorbed, and this multiplied by 100 and divided by the weight of fat taken is the iodine absorption.

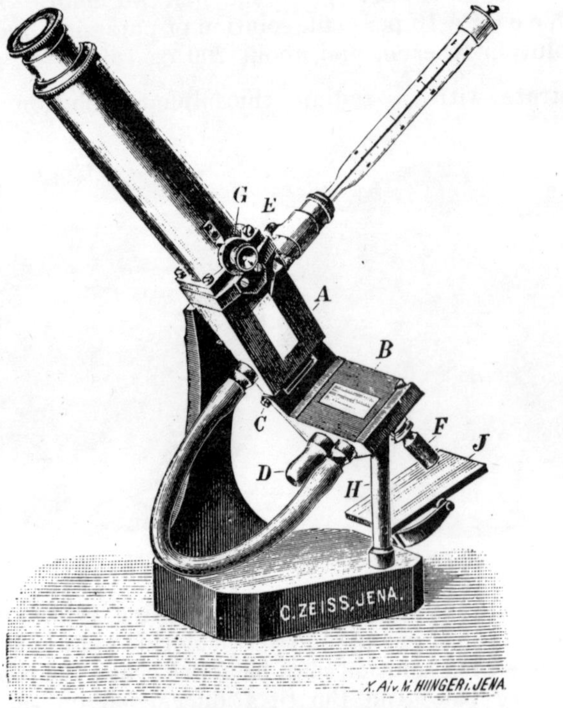

Fig. 36.—Refractometer.

Determination of Refractive Index.—The Zeiss butyro-refractometer is employed for the determination of the refractive index ; it consists of two water-jacketed prisms, between which the substance is placed, a mirror to reflect light through these, an eye-piece with a scale in it by which the refractive index is read off, and a thermometer for observing the temperature. An apparatus for providing a stream of water of constant temperature can be used, or in default of this stream of water warmed to the required degree can be run through the jacket by india-rubber tubes from any fair-sized vessel; though it is an advantage to use water at a constant temperature, the cooling of a large bulk of water is sufficiently slow to keep the prisms at practically a constant temperature during the reading.

To make an observation, take the refractometer (Fig. 36) out of its case, stand it up, screw in the thermometer, and connect the india-rubber tubes carrying the water to the inlet and outlet tubes of the jacket. Turn the milled head, and open and throw back on to its support the lower prism; see that both glass surfaces are clean (a cloth dipped in alcohol, followed by a dry soft cloth is best for cleaning), and place a drop of the melted fat in the centre of the surface of the lower prism (Fig. 37), and close it.

Arrange the mirror to reflect either daylight or the light of a lamp through the prisms, and observe the point on the scale where the dark shadow comes, focusing the eye-piece if necessary; make two or three observations till the position of the shadow is constant, and read the temperature at the same time. The thermometer may be graduated either in Centigrade degrees, or in the normal reading for butter fat at the temperature of the water; if of the former kind, both readings are noted, and the refractometer readings corrected to a standard temperature by multiplying the difference between the observed temperature and the standard temperature by 0·55, and adding or subtracting the result, as the temperature is higher or lower than the standard, to the refractometer reading. The correction of the refractometer reading for temperature may be made by the dairy scale. If the thermometer be of the latter type, the observed

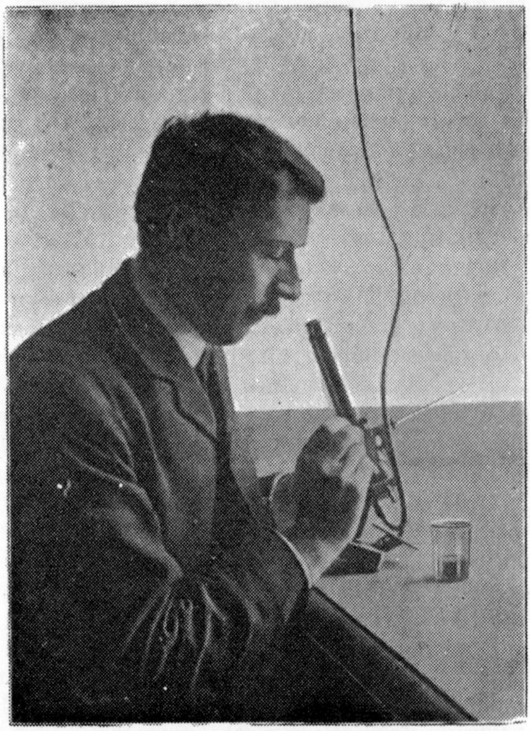

Fig. 37.—Using Refractometer.

reading of the refractometer is subtracted from the reading of the thermometer, and the result is expressed in degrees less than the standard.

Table VIII. gives the standard readings for each five degrees Centigrade.

TABLE VIII.

Temperature.	Scale Divisions.
25	52·5
30	49·8
35	47·0
40	44·2
45	41·5

The average figure for butter is 46° at 35° C. or a little below, and margarine gives about 54°. Genuine butter sometimes gives a reading higher than 47°, and the limits found are 43·5° to 49°.

A standard solution (*normal flüssigkeit*) is supplied with the instrument, and the scale should be adjusted from time to time with this; a point is marked on the scale where the standard solution should read, and a key is provided for adjusting the scale to this.

Detection of Sesamé Oil.—Add to 10 c.c. of the melted butter 0·1 c.c. of a 2 per cent. alcoholic solution of furfural, add 1 c.c. of strong hydrochloric acid, shake well and add 10 c.c. of chloroform. A crimson coloration of the aqueous layer indicates sesamé oil. Sometimes coal-tar colours are added to the butter which give a red colour with hydrochloric acid alone; make sure of the absence of these by testing a little of the butter with hydrochloric acid. If these be present use the fatty acids (p. 82) for the test.

If furfural be not available dissolve 0·1 gramme of sugar in 5 c.c. of hydrochloric acid, and shake with 10 c.c. of melted fat. This will also give a crimson colour with sesamé oil.

Detection of Coco-nut Oil—*Hinks' Method.*—Dissolve 5 c.c. of fat in 10 c.c. of ether, in a corked test-tube, and pack this in ice; after thirty minutes pour the whole mass on a filter, and evaporate the filtrate; take up the fat with 3 to 4 c.c. of 96 to 97 per cent. alcohol, boil when the fat should all dissolve, and cool to 5° C. for fifteen minutes, and filter this; cool the filtrate to 0° C. for two to three hours. Place a little of the cooled solution on a well-cooled slide, and examine rapidly with a $\frac{1}{2}$ or $\frac{2}{3}$ inch objective. Butter crystallises in round globular masses, while coco-nut oil gives characteristic feathery crystals, easily recognised after a little practice; other fats such as lard give somewhat similar crystals.

The Application of Analysis to the Solution of Problems—*The Detection of Adulteration.*—The President of the Board of Agriculture, under his powers authorised by Clause 4 of the Sale of Food and Drugs Act, 1899, has laid down the limit of 16 per cent. of water in butter, and any quantity above this amount is presumed to have been added. Occasionally water has been worked into butter to add to its weight, but a more frequent occurrence of excess of water occurs in milk-blended butter; in the latter the "curd" will be in the proportion of 1 part for each 10 parts of water, in the former much less.

A washed butter may be distinguished from an unwashed one by containing less than 1 part of curd to 10 parts of water, usually only about 0·5 part; unwashed butter contains about 1 part; a larger percentage of curd indicates that the butter was churned from very sour cream, and will not keep well, or that it has been adulterated with casein. A higher percentage of casein than 0·5 indicates the presence of added casein; this form of adulteration is not now common. Cane-sugar or honey is occasionally added; the test given on p. 62 applied to the aqueous portion obtained on melting out will detect this. The percentage of salt should not exceed 3 in a mild salt butter, but may go up to

5 per cent. or more in other samples. Irish pickled butters are high in water and in salt.

The Preservatives Committee of the Local Government Board recommended that no preservative except boric acid be allowed in butter, and that only to the extent of 0·5 per cent.; this has not yet been legalised.

The detection of margarine is more difficult; the Butter Regulations Committee of the Board of Agriculture has recommended a limit for the Reichert-Wollny figure of 24 c.c., but their recommendation has not yet been adopted.

This figure, as well as 89 as a superior limit for insoluble fatty acids, 5 per cent. as the minimum for soluble fatty acids, 220 for the potash required for saponification, 0·910 for the density, 42 per cent. for the iodine absorption, and 48° as a maximum Zeiss butyro-refractometer figure at 35° C., may be taken as indicating the border-line of genuine butters. Certain butters, however, especially those prepared from the milk of cows exposed to cold climates and near the end of lactation, yield figures beyond these limits. The appearance on melting, and under polarised light, may be useful as confirmatory, and a reaction for sesamé oil will establish the presence of margarine, while vegetable oils give the phytosteryl acetate test.

The average figures for butter and margarine are:—

TABLE IX.

	Butter.	Margarine.	Coco-nut Oil.
Reichert-Wollny figure,	29 c.c.	Practically none	8
Polenske figure, . .	2·5	1·0	17
Insoluble fatty acids, .	87·5 %	95·5 %	82
Soluble fatty acids, .	6·0 %	None	?
Potash required, . .	227 %	195 %	258
Density, ...	0·913	0·902	?
Iodine absorption, .	37	55	9
Refractometer at 35°, .	46°	54°	43°

Coco-nut oil has a different composition from that of margarine; but as it lowers the Reichert-Wollny figure, while at the same time raising the Polenske figure and potash absorption, and lowering the insoluble fatty acids, iodine absorption and the refractometer figure, its detection is not difficult. Positive evidence of its presence can be obtained by Hinks' method.

Should the Polenske figure exceed the maximum calculated from the Reichert-Wollny and Kirschner figures, the presence of coco-nut oil may be considered to be proved, and the percentage may be calculated by the formula $C = \frac{P - P'}{14 \cdot 4} \times 100$, where P' = mean Polenske figure calculated equivalent to the Reichert-Wollny figure found plus half the Polenske figure.

Under the Margarine Act, margarine containing more than 10 per cent. of butter is not permitted. The estimation of small amounts of butter can be performed with great accuracy from the Kirschner and Polenske figures; the percentage of butter is given by the formula

$$B = \frac{K - (0 \cdot 262\, P^{0 \cdot 63} + 0 \cdot 09)}{0 \cdot 242}.$$

The value of $0 \cdot 262\, P^{0 \cdot 63} + 0 \cdot 09$ is given on the dairy scale or in the table below.

P.	Value.	P.	Value.
0·5	0·26	8	1·06
1	0·35	9	1·13
2	0·50	10	1·21
3	0·61	11	1·28
4	0·72	12	1·34
5	0·81	13	1·41
6	0·90	14	1·47
7	0·98	15	1·53

Application of Analysis to Buttermaking.—The acidity of the cream should be taken to see that it is properly ripened; an acidity of 60° to 70° is usually suitable.

The fat in the butter-milk is an important item, as this represents loss; 0·2 per cent. shows satisfactory churning, while more than 1 per cent. even with very thick cream indicates bad working; either the ripening has not been properly done, the churning is too rapid, or the temperatures are wrong.

The amount of butter yielded by milk may be calculated by subtracting 0·1 from the percentage of fat, multiplying by 7, and dividing by 60; this gives the number of pounds of butter per gallon of milk.

The following Table will give the amount of butter that may be expected from cream of any percentage of fat:—

TABLE X.

Percentage of Fat in Cream.	Quarts of Cream Churned.								
	1	2	3	4	5	6	7	8	9	10
15	0·44	0·87	1·31	1·74	2·18	2·61	3·05	3·48	3·92	4·35
16	0·47	0·93	1·40	1·86	2·33	2·79	3·26	3·72	4·19	4·65
17	0·50	0·99	1·49	1·98	2·48	2·98	3·47	3·97	4·46	4·96
18	0·53	1·05	1·58	2·10	2·63	3·16	3·68	4·21	4·73	5·26
19	0·56	1·11	1·67	2·23	2·79	3·34	3·90	4·46	5·01	5·57
20	0·59	1·17	1·76	2·35	2·94	3·52	4·11	4·70	5·28	5·87
21	0·62	1·23	1·85	2·46	3·08	3·70	4·31	4·93	5·54	6·16
22	0·65	1·29	1·94	2·58	3·23	3·87	4·52	5·16	5·81	6·45
23	0·68	1·35	2·03	2·70	3·38	4·05	4·73	5·40	6·08	6·75
24	0·70	1·41	2·11	2·82	3·52	4·22	4·93	5·63	6·34	7·04
25	0·73	1·47	2·20	2·93	3·67	4·40	5·13	5·86	6·60	7·33
26	0·76	1·52	2·29	3·05	3·81	4·57	5·33	6·10	6·86	7·62
27	0·79	1·58	2·37	3·16	3·96	4·75	5·54	6·33	7·12	7·91
28	0·82	1·64	2·46	3·28	4·10	4·91	5·73	6·55	7·37	8·19
29	0·85	1·70	2·54	3·39	4·24	5·09	5·94	6·78	7·63	8·48
30	0·88	1·75	2·63	3·51	4·38	5·26	6·14	7·02	7·89	8·77
31	0·91	1·81	2·72	3·62	4·53	5·43	6·34	7·24	8·15	9·05
32	0·93	1·87	2·80	3·74	4·67	5·60	6·54	7·47	8·41	9·34
33	0·96	1·92	2·89	3·85	4·81	5·77	6·73	7·70	8·66	9·62
34	0·99	1·98	2·97	3·96	4·96	5·95	6·94	7·93	8·92	9·91
35	1·02	2·04	3·06	4·08	5·10	6·11	7·13	8·15	9·17	10·19
36	1·05	2·09	3·14	4·18	5·23	6·28	7·32	8·37	9·41	10·46
37	1·07	2·15	3·22	4·30	5·37	6·44	7·52	8·59	9·67	10·74
38	1·10	2·20	3·30	4·40	5·51	6·61	7·71	8·81	9·91	11·01
39	1·13	2·26	3·39	4·52	5·65	6·77	7·90	9·03	10·16	11·29
40	1·16	2·31	3·47	4·62	5·78	6·94	8·09	9·25	10·40	11·56
41	1·18	2·37	3·55	4·73	5·92	7·10	8·28	9·46	10·65	11·83
42	1·21	2·42	3·63	4·84	6·05	7·25	8·46	9·67	10·88	12·09
43	1·24	2·47	3·71	4·94	6·18	7·42	8·65	9·89	11·12	12·36
44	1·26	2·52	3·79	5·05	6·31	7·57	8·83	10·10	11·36	12·62
45	1·29	2·58	3·87	5·16	6·45	7·73	9·02	10·31	11·60	12·89
46	1·32	2·63	3·95	5·26	6·58	7·89	9·21	10·52	11·84	13·15
47	1·34	2·68	4·02	5·36	6·70	8·04	9·38	10·72	12·06	13·40
48	1·36	2·73	4·09	5·46	6·82	8·18	9·55	10·91	12·28	13·64
49	1·39	2·77	4·16	5·55	6·94	8·32	9·71	11·10	12·48	13·87
50	1·41	2·82	4·23	5·64	7·05	8·46	9·87	11·28	12·69	14·10

The calculation of yields of butter from either milk or cream may be performed by the dairy scale.

CHAPTER VI.

THE ANALYSIS OF CHEESE.

Estimation of Water, Ash, and Salt.—Place 2 or 3 grammes of cheese cut into small pieces in a small flat-bottomed basin, and keep in the water-oven for six hours; the drying proceeds better if the basin be inclined so that the fat runs off the drying cheese; weigh and return to the water-oven, and weigh again at intervals of one hour till the loss is less than 1 milligramme per hour; the loss may be taken as water.

Pour off as much fat as possible, and macerate the residue in hot amyl alcohol; pour off the amyl alcohol as completely as possible, and ignite the residue as in determining the ash of milk (p. 12); to determine the salt make a determination of chlorine as on p. 14; each cubic centimetre of $\frac{N}{10}$ silver nitrate is equal to 0·00585 gramme salt.

Estimation of Fat.—Weigh about 2 grammes of cheese, cut into small pieces, and transfer to a Stokes tube, add 8 c.c. of water, and heat gently till the cheese is softened and disintegrated; then add 10 c.c. of hydrochloric acid, and treat as in the Werner-Schmid method (p. 30).

Estimation of Total Nitrogen.—Weigh about 1 gramme of cheese and treat by the Kjeldahl method (p. 36). This multiplied by 6·39 will give the protein.

Estimation of Products of Ripening.—Weigh

10 grammes of cheese, place in a small mortar, and add 25 c.c. of boiling water; with a pestle grind up the cheese and water, and pour off the solution through a filter, collecting the filtrate in a 250 c.c. flask; repeat the treatment with 25 c.c. of boiling water till nine portions have been used; cool the total filtrates, make up to 250 c.c. and mix well. Evaporate 50 c.c. in a weighed basin, and weigh the solids after drying till the loss is less than 1 milligramme per hour; ignite, and weigh the ash; the weight of the solids less that of the ash represents the products of ripening. The difference between 100, and the sum of the water, fat, ash, and products of ripening, may be taken as unaltered para-casein.

The products of ripening may be differentiated; to 50 c.c. of the filtrate add 5 c.c. of copper sulphate, and treat as in the Ritthausen method for the estimation of proteins (p. 34); the proteins estimated in this way may be termed primary products of ripening, and the remainder secondary products.

Examination of the Fat.—Dry 25 to 50 grammes of cheese till the fat runs out; extract with ether, and wash the ethereal solution with water in a separating funnel; remove the ethereal layer, drive off the ether, and dry the fat till the ether is removed completely. Examine the fat as directed for butter fat; usually a Reichert-Wollny figure is sufficient.

The Application of Analysis to the Solution of Problems—*Detection of Adulteration.*—Practically the only adulterations of cheese consist in the removal of fat from the milk before curdling, and the addition of foreign matter.

The removal of fat may be judged if the fat is less than 45 per cent. of the dried cheese, or if the fat is less than six times the total nitrogen, both of which standards lead to practically the same result; a large number of cheeses are made with half-skimmed milk (*e.g.*, the evening's milk is skimmed, and mixed with the fresh morning's milk); these fail to comply with the above standards, and should be sold as half-skim cheeses; other cheeses are made from skim-milk alone—*e.g.*, Dutch cheese (though cheeses are made in Holland with whole milk also).

Another method, particularly applicable to cream cheese, is to deduce the composition of the cream (or milk) used for the preparation of the sample, as follows:—

$$\text{Fat in original milk} = \frac{100\,F}{35{\cdot}4\,P + F} + 0{\cdot}25.$$

$$\text{Solids not fat} \quad ,, \quad = \frac{333\,P}{35{\cdot}4\,P + F}.$$

F = fat, P = protein (nitrogen $\times$ 6·39).

The quality of the cheese is then judged by the quality of the milk or cream used for its preparation.

The detection of "margarine-cheese" is accomplished by the examination of the fat; practically the same standards as for butter may be used; it must be remembered, however, that during the ripening the fat is slightly attacked, and the percentage of volatile acids may be somewhat lowered by this cause.

Preservatives need not be looked for in cheese.

Application of Analysis to Cheese-making—*Estimation of the Curd by Lindet's Method.*—Estimate the fat and specific gravity as previously described; to 100 c.c. of milk add 0·01 gramme of rennet powder, and keep at 42° C. till curdled; cut up the curd and allow it to settle, and strain off the whey through muslin; cool the whey to 15·5° C.

and estimate the specific gravity and the fat as before.

Add the degrees of gravity and the percentage of fat of the milk and subtract the sum of the degrees of gravity and the percentage of fat of the whey; the difference divided by 3·5 will give the percentage of dry curd available for cheese-making.

In practice the whey obtained during cheese-making may be tested instead of a separate preparation of whey being made; the sample for testing should, however, be removed as early as possible.

The difference in acidity (*see* p. 15) between the milk and the whey divided by 3·5 will also give a rough estimation of dry curd.

This will give an idea of the value of milk for cheese-making. The percentage of dry curd plus the percentage of fat less 0·25 divided by 0·055 will give the number of pounds of cheese per 100 gallons of milk.*

A fermentation test is useful; plug a number of clean test-tubes with cotton-wool, and sterilise by heating to 150° C. in an air-bath for half an hour. Place 10 c.c. of the milk to be examined in one of these, and keep it at blood-heat for eighteen hours; if the precipitated curd be distended by bubbles of gas the milk will not make good cheese.

The acidity of the milk before renneting should be estimated (p. 15), and that of the whey after the curd is cut, and the whey running from the curd at intervals. As an example of the use of the acidity test, in Cheddar cheese-making the best acidity of the milk for renneting is 22°-24°; the acidity of the whey is less than this, but constantly increases;

* The factor 0·055 will require modification according to the kind of cheese made; it applies fairly well to Cheddar and Cheshire cheese.

the whey should be drawn at about 22° acidity, and the curd vatted when the whey draining off has an acidity of about 100°.

The acidity of the curd may also be judged by the hot-iron test; an iron is heated and allowed to cool till it can just be touched with the finger; the acidity of the curd is judged by the length of the string which is formed when the iron is pressed on the curd and withdrawn.

The strength of rennet is determined by weighing out 0·5 gramme of a solid extract, or measuring 5 c.c. of a liquid extract, and diluting to 100 c.c. 1 c.c. of the solution is added to 100 c.c. of separated milk of acidity 20° at a temperature of 35° C.; the temperature is kept at 35°, and the milk slowly stirred with a thermometer till it curdles, which is indicated by the path of the thermometer becoming visible; the time which has elapsed since the addition of the rennet is noted, and the strength of the rennet calculated as parts of rennet which will cause curdling in forty minutes by the following formulæ:—

$$\text{Strength} = \frac{800{,}000}{T} \text{ for solid extracts,}$$

$$\text{or} \quad \frac{80{,}000}{T} \text{ for liquid extracts,}$$

where T = time in minutes.

The curdling should not take less than five minutes nor more than ten minutes, and should this not be the case less or more of the rennet solution should be used, and the results increased or decreased proportionately.

TABLE XI.

For Correcting Specific Gravity to 60° F.

(*see* p. 9).

Temperature. Degrees F.	Degrees of Specific Gravity observed.											
	25	**26**	**27**	**28**	**29**	**30**	**31**	**32**	**33**	**34**	**35**	**36**
	Specific Gravity corrected to 60° F.											
40	23·5	24·5	25·5	26·4	27·3	28·2	29·1	30·0	31·0	31·9	32·8	33·7
42	23·6	24·6	25·6	26·5	27·5	28·4	29·3	30·2	31·1	32·0	32·9	33·9
44	23·8	24·8	25·8	26·7	27·7	28·6	29·5	30·4	31·3	32·2	33·1	34·1
46	23·9	24·9	25·9	26·8	27·8	28·7	29·6	30·5	31·4	32·4	33·3	34·3
48	24·0	25·0	26·0	26·9	27·9	28·8	29·7	30·6	31·6	32·6	33·5	34·5
50	24·1	25·1	26·1	27·0	28·0	29·0	29·9	30·9	31·8	32·8	33·7	34·7
52	24·3	25·2	26·2	27·2	28·1	29·1	30·1	31·1	32·0	33·0	33·9	34·9
54	24·5	25·4	26·4	27·4	28·4	29·3	30·3	31·3	32·3	33·3	34·2	35·1
56	24·6	25·6	26·6	27·6	28·6	29·6	30·5	31·5	32·5	33·5	34·4	35·4
58	24·8	25·8	26·8	27·8	28·8	29·8	30·8	31·7	32·7	33·7	34·7	35·7
60	**25·0**	**26·0**	**27·0**	**28·0**	**29·0**	**30·0**	**31·0**	**32·0**	**33·0**	**34·0**	**35·0**	**36·0**
62	25·2	26·2	27·3	28·3	29·3	30·3	31·3	32·3	33·3	34·3	35·3	...
64	25·4	26·5	27·5	28·5	29·5	30·5	31·5	32·6	33·6	34·6	35·6	...
66	25·6	26·7	27·7	28·7	29·8	30·8	31·8	32·9	33·9	34·9	35·9	...
68	25·9	27·0	28·0	29·0	30·1	31·1	32·1	33·2	34·2	35·2	36·2	...
70	26·1	27·2	28·2	29·2	30·3	31·3	32·4	33·4	34·5	35·5	36·5	...
72	26·4	27·4	28·4	29·5	30·5	31·6	32·6	33·7	34·7	35·8	...	...
74	26·6	27·7	28·7	29·7	30·8	31·9	32·9	34·0	35·0	36·1	...	...
76	26·9	27·9	28·9	29·9	31·0	32·2	33·3	34·4	35·4	36·5	...	...
78	27·2	28·2	29·2	30·3	31·4	32·5	33·6	34·7	35·8	36·9	...	...
80	27·4	28·4	29·5	30·6	31·7	32·8	33·9	35·0	36·1	...	...	...

TABLE XII.

For Calculating Solids not Fat from Fat and Specific Gravity (*see* p. 43).

Correction.	* Specific Gravity (Degrees).							
− 1·0	25·0	25·5	26·0	26·5	27·0	27·5	28·0	28·5
*	**29·0**	**29·5**	**30·0**	**30·5**	**31·0**	**31·5**	**32·0**	**32·5**
+ 1·0	33·0	33·5	34·0	34·5	35·0	35·5	36·0	36·5
Fat.	Solids not Fat.							
0·0	7·40	7·50	7·65	7·75	7·90	8·00	8·15	8·25
0·25	7·45	7·55	7·70	7·80	7·95	8·05	8·20	8·30
0·5	7·50	7·60	7·75	7·85	8·00	8·10	8·25	8·35
0·75	7·55	7·65	7·80	7·90	8·05	8·15	8·30	8·40
1·0	7·60	7·70	7·85	7·95	8·10	8·20	8·35	8·45
1·25	7·65	7·75	7·90	8·00	8·15	8·25	8·40	8·50
1·5	7·70	7·80	7·95	8·05	8·20	8·30	8·45	8·55
1·75	7·75	7·85	8·00	8·10	8·25	8·35	8·50	8·60
2·0	7·80	7·90	8·05	8·15	8·30	8·40	8·55	8·65
2·75	7·85	7·95	8·10	8·20	8·35	8·45	8·60	8·70
2·5	7·90	8·00	8·15	8·25	8·40	8·50	8·65	8·75
2·75	7·95	8·05	8·20	8·30	8·45	8·55	8·70	8·80
3·0	8·00	8·10	8·25	8·35	8·50	8·60	8·75	8·85
3·25	8·05	8·15	8·30	8·40	8·55	8·65	8·80	8·90
3·5	8·10	8·20	8·35	8·45	8·60	8·70	8·85	8·95
3·75	8·15	8·25	8·40	8·50	8·65	8·75	8·90	9·00
4·0	8·20	8·30	8·45	8·55	8·70	8·80	8·95	9·05
4·25	8·25	8·35	8·50	8·60	8·75	8·85	9·00	9·10
4·5	8·30	8·40	8·55	8·65	8·80	8·90	9·05	9·15
4·75	8·35	8·45	8·60	8·70	8·85	8·95	9·10	9·20
5·0	8·40	8·50	8·65	8·75	8·90	9·00	9·15	9·25
Change at†	0·2	0·1	0·2	0·1	0·2	0·1	0·2	0·1

* The solids not fat are correct for the middle line of specific gravity; if the specific gravity falls in the top line subtract 1 from the solids not fat; thus 26·0 sp. gr. and 3·0 % fat give 7·25 % solids not fat; if the specific gravity falls in the bottom line, add 1 to the solids not fat; thus 34·0 sp. gr. and 3·0 % fat give 9·25 % solids not fat.

† The last line indicates where the change of solids not fat takes place; in a column with 0·2 at the bottom use the column itself for the percentage of fat given, and 0·05, 0·10, and 0·15 above, but use the figure immediately below for 0·2 above; thus 30·0 sp. gr. and 3·1 % fat give 8·25 % solids not fat, but 30·0 sp. gr. and 3·2 % fat give 8·30 % solids not fat; in a column with 0·1 at the bottom, use the column itself for the percentage of fat given and 0·05 above, but change to the figure immediately below for 0·1 or more above; thus 30·5 sp. gr. and 3·30 % fat give 8·40 % solids not fat, but 30·5 sp. gr. and 3·40 % fat give 8·45 % solids not fat.

TABLE XIII.

For Calculating Fat from Total Solids and Specific Gravity (*see* p. 43).

Specific Gravity Observed.										
	A25·0 B29·0 C33·0	25·5 29·5 33·5	26·0 30·0 34·0	26·5 30·5 34·5		A27·0 B31·0 C35·0	27·5 31·5 35·5	28·0 32·0 36·0	28·5 32·5 36·5	
Total Solids A	Fat.				Total Solids B	Fat.				Total Solids C
7·00	0·50	0·40	0·30	0·20	8·00	0·10	...	...	...	9·00
7·25	0·70	0·60	0·50	0·40	8·25	0·30	0·20	0·10	...	9·25
7·50	0·90	0·80	0·70	0·60	8·50	0·50	0·40	0·30	0·20	9·50
7·75	1·15	1·05	0·90	0·80	8·75	0·70	0·60	0·50	0·40	9·75
8·00	1·35	1·25	1·15	1·05	9·00	0·90	0·80	0·70	0·60	10·00
8·25	1·55	1·45	1·35	1·25	9·25	1·15	1·05	0·90	0·80	10·25
8·50	1·75	1·65	1·55	1·45	9·50	1·35	1·25	1·15	1·05	10·50
8·75	1·95	1·85	1·75	1·65	9·75	1·55	1·45	1·35	1·25	10·75
9·00	2·15	2·05	1·95	1·85	10·00	1·75	1·65	1·55	1·45	11·00
9·25	2·40	2·30	2·15	2·05	10·25	1·95	1·85	1·75	1·65	11·25
9·50	2·60	2·50	2·40	2·30	10·55	2·15	2·05	1·95	1·85	11·50
9·75	2·80	2·70	2·60	2·50	10·75	2·40	2·30	2·15	2·10	11·75
10·00	3·00	2·90	2·80	2·70	11·00	2·60	2·50	2·40	2·30	12·00
10·25	3·20	3·10	3·00	2·90	11·25	2·80	2·70	2·60	2·50	12·25
10·50	3·40	3·30	3·20	3·10	11·50	3·00	2·90	2·80	2·70	12·50
10·75	3·65	3·55	3·40	3·30	11·75	3·20	3·10	3·00	2·90	12·75
11·00	3·85	3·75	3·65	3·55	12·00	3·40	3·30	3·20	3·10	13·00
11·25	4·05	3·95	3·85	3·75	12·25	3·65	3·55	3·40	3·30	13·25
11·50	4·25	4·15	4·05	3·95	12·50	3·85	3·75	3·65	3·55	13·50
11·75	4·45	4·35	4·25	4·15	12·75	4·05	3·95	3·85	3·75	13·75
12·00	4·65	4·55	4·45	4·35	13·00	4·25	4·15	4·05	3·95	14·00
12·25	4·90	4·80	4·65	4·55	13 25	4·45	4·35	4·25	4·15	14·25
12·50	5·10	5·00	4·90	4·80	13·50	4·65	4·55	4·45	4·35	14·50
12·75	5·30	5·20	5·10	5·00	13·75	4·90	4·80	4·65	4·55	14·75
13·00	5·50	5·40	5·30	5·20	14·00	5·10	5·00	4·90	4·80	15·00

If the total solids do not exactly agree with a figure in the Table add the excess of total solids to the fat; thus, given 30·5 sp. gr. and 11·8 % total solids, add 0·05 to fat corresponding with 11·75 % total solids = 3·35 % fat.

If the specific gravity lies in line marked A use the total solids column marked A; if in B use total solids column B; if in C use total solids column C.

APPENDIX.

PREPARATION OF STANDARD SOLUTIONS.

As the preparation of a standard solution is rarely, if ever, undertaken while an analysis is being performed, and as also it is tedious to wade through a description of a method when seeking only for the quantities to be weighed or measured for the solution, details of the preparation have been removed from the text; the arrangement is alphabetical.

Alcoholic Soda Solution.—$\frac{N}{2}$ approximately. Add 25 c.c. of a 50 per cent. caustic soda solution (*see* p. 106) to 1 litre of alcohol (sp. gr. 0·830) and mix well; allow the solution to stand a day, filter into a dry stoppered bottle, and titrate (*see* Titration) against $\frac{N}{2}$ hydrochloric acid (*see* p. 104) to ascertain strength. As this solution loses strength on keeping, it must be titrated each time it is used.

Amyl Alcohol (for Gerber test) should conform to the following:—Density, 0·8145 to 0·816 at 15·5° C. (Water at 15·5° C. = 1.) Boiling-point: should not begin to boil below 124° C.; not more than 5 c.c. should distil at 127·5° C. when 25 c.c. are boiled in a 100 c.c. flask, and the boiling-point should not rise above 132° C.; 10 c.c. should mix completely with 10 c.c. of hydrochloric acid (sp. gr. 1·17) and the addition of 1·5 c.c. of water should produce a permanent turbidity. 2 c.c. treated in the Gerber bottle with 10 c.c. water and 10 c.c. sulphuric acid should yield no layer of "fat."

Barium Chloride Solution, $\frac{N}{5}$.—24·45 grammes of crystallised barium chloride are dissolved in 1 litre of distilled water.

Baryta Solution.—Dissolve 33 to 35 grammes of barium hydroxide in 2 litres of water; allow to settle, and store in a bottle fitted with a soda lime tube to prevent access of CO_2. Check the strength by titration (*see* p. 107).

Caustic Soda.—*See* Soda.

Fehling's Solution.—Copper sulphate solution:—Dissolve 34·639 grammes of crystallised copper sulphate in water, and make up to 500 c.c. Alkaline tartrate solution:—Dissolve 173 grammes of pure sodium potassium tartrate (Rochelle Salt), and 51 to 55 grammes of sodium hydroxide of good quality in water, and make up to 500 c.c.

Hydrochloric Acid $\frac{N}{2}$.—Dilute 50 c.c. strong hydrochloric acid to 1 litre; ignite a quantity of pure sodium bicarbonate at a dull red heat, and cool in a desiccator; weigh (accurately to 1 milligramme) portions of this, of about 1 gramme each, dissolve in water, add a little methyl red, and run in the hydrochloric acid solution from a burette till the sodium carbonate solution turns pink; the solution should be covered as much as possible with a watch-glass to prevent loss by spirting as the CO_2 is given off. Wash the watch-glass into the solution, and boil it, cool, and continue the titration till the pink colour is permanent after boiling. If the solution is $\frac{N}{2}$, 20 c.c. should be required for each 0·53 gramme of sodium carbonate; if this quantity be not used dilute in the proportion of the quantity used to

20 c.c., or use the solution as it is, and multiply by 20 divided by the quantity used to obtain its value in terms of $\frac{N}{2}$ acid. $\frac{N}{10}$ solution is made from $\frac{N}{2}$ solution by diluting 200 c.c. to 1 litre with well-boiled water.

Iodine Solution (Wijs).—Dissolve 13 grammes of iodine in 1 litre of 99 per cent. glacial acetic acid; pass in a current of chlorine till the titre of the solution has been doubled; this point is indicated by a change of colour. Add a little more iodine to prevent the formation of iodine trichloride.

$\frac{N}{10}$.—Dissolve 20 grammes of potassium iodide in 10 c.c. (not more) of warm water; add 12·69 grammes of resublimed iodine, and when all is dissolved dilute to 1 litre.

Magnesia Mixture.—Mix 60 grammes magnesium chloride ($MgCl_2 6OH_2$), 145 grammes ammonium chloride, 600 c.c. water, and 300 c.c. ammonia solution (sp. gr. 0·880).

Mercuric Nitrate Solution (Wiley).—Dissolve mercury in twice its weight of nitric acid (sp. gr. 1·42), and, after solution, add an equal bulk of water.

Mayer's Reagent.—Dissolve 13·546 grammes of mercuric chloride and 49·8 grammes of potassium iodide in water and dilute to 1 litre.

Nessler Solution.—Dissolve 17·5 grammes of potassium iodide in 100 c.c. of water, next dissolve 15 grammes of mercuric chloride in 300 c.c. of water, and mix the two solutions, wash the heavy precipitate that forms well by decantation, and dissolve it in 17·5 grammes of potassium iodide in 100 c.c. of water, add a few drops of mercuric chloride solution till a red precipitate, insoluble on shaking, is produced, and dilute to about 500 c.c., cooling in ice water, and mix with so much of a 50 per cent.

caustic soda solution as is equal to 105 grammes of sodium hydroxide previously diluted with 200 c.c. of water and cooled in ice water. Cool well during mixing, and make up to 1 litre. The solution is left to settle and the clear portion decanted for use.

Phenolphthalein Solution.—Dissolve 5 grammes of phenolphthalein in 600 c.c. of alcohol and dilute to 1 litre with water. Methylated spirit may be used, and if necessary the solution should be treated with animal charcoal, and filtered till clear.

Silver Nitrate, $\frac{N}{10}$.—Dissolve 17·000 grammes of silver nitrate in 1 litre of water.

Soda Solution, 50 per cent.—Dissolve 250 grammes of caustic soda (purified by alcohol) in 250 c.c. of water; allow to stand till clear, and store in the apparatus described on p. 77.

For Kjeldahl Process.—Dissolve 300 grammes of caustic soda in water and make up to 1 litre. After standing a few days, filter through glass wool.

$\frac{N}{2}$.—Dilute 25 c.c. of the 50 per cent. solution to 1 litre. Check the strength by titration (*see below*).

$\frac{N}{10}$.—Dilute 5 c.c. of the 50 per cent. solution to 1 litre. Check the strength by titration.

Sodium Hypobromite Solution.—Add 1 c.c. of bromine to 10 c.c. of soda solution (for Kjeldahl process).

Strontia Solution.—Dissolve 28 to 30 grammes of strontium hydroxide in 2 litres of water. Treat as baryta solution (p. 104).

Sulphuric Acid—*For Gerber Process.*—Should contain 90 to 91 per cent. H_2SO_4. Acid of specific gravity 1·820 to 1·825 fulfils this requirement.

For Reichert Process.—Dilute 25 c.c. strong sulphuric acid to 1 litre of water; 2 c.c. of 50 per cent. soda solution should neutralise about 35 c.c. of this solution; adjust the strength, if necessary, by adding acid or water.

$\frac{N}{1}$.—Dilute 28 c.c. of strong sulphuric acid to 1 litre. Check the strength as for hydrochloric acid.

$\frac{N}{10}$.—Dilute 100 c.c. of $\frac{N}{1}$ acid to 1 litre.

Thiosulphate Solution.—Dissolve 25 grammes of pure sodium thiosulphate and 1 gramme of salicylic acid in 1 litre of water. Allow to stand for a few days and filter. Weigh accurately about 0·25 gramme of iodine in a small stoppered flask, add 2 grammes potassium iodide and 2 c.c. of water, and shake gently till the iodine is dissolved. Dilute with water and transfer to a larger vessel, and run in the thiosulphate solution from a burette till the yellow colour just disappears. Repeat this two or three times, and from the mean results calculate the weight of iodine equivalent to 1 c.c.

Titration.—Take two burettes, fill one with the alkaline solution, and the other with hydrochloric acid solution of corresponding strength. Measure 25 c.c. of the acid solution, add a few drops of phenolphthalein solution (or cochineal or methyl red solution if this be used in the experiment for which the alkali is to be employed), and run in the alkali till a pink colour (violet with cochineal, yellow with methyl red) is produced. Note the volume of alkali used. Repeat this experiment two or three times, and from the mean of the results calculate the ratio of the alkali solution to the acid solution. Thus, if 24·2, 24·15, and 24·2 c.c. of $\frac{N}{2}$

soda were used for 25 c.c. acid the ratio is $\frac{25}{24 \cdot 17}$ $= 1 \cdot 034$ or 1 c.c. of soda $= 1 \cdot 034$ c.c. acid.

If the acid is strictly $\frac{N}{2}$ the soda is $\frac{N}{2} \times 1 \cdot 034$; if the acid be, for instance, $\frac{N}{2} \times 1 \cdot 011$, the soda is then $\frac{N}{2} \times 1 \cdot 034 \times 1 \cdot 011 = \frac{N}{2} \times 1 \cdot 0454$.

The soda may be diluted in the ratio of 1,000 parts to 1,045·4 parts—*i.e.*, 45·4 c.c. of water are added to each litre, but it is preferable to use the solution as it is, and multiply the results by the factor.

To standardise $\frac{N}{10}$ solutions $\left(\frac{N}{11}\text{ strontia}\right)$ 5 c.c. of the $\frac{N}{2}$ hydrochloric acid should be carefully measured, a few drops of phenolphthalein solution added, and the alkali run in till a faint pink colour is seen. Several titrations should be made, and the mean taken. This figure divided into 2·5 will give the strength of the solution in terms of normal. Thus, if 26·2, 26·3, and 26·25 c.c. were used in three experiments the mean is 26·25, and the strength is $\frac{2 \cdot 5}{26 \cdot 25} = 0 \cdot 09505$ N.

ADDENDUM.

VOLUMETRIC ESTIMATION OF MILK-SUGAR.

10 c.c. of milk are weighed into a 100 c.c. flask and diluted with 50 c.c. of distilled water; 10 c.c. of Mayer's reagent (see Appendix) and 2 c.c. of $\frac{N}{1}$ sulphuric acid are added, the whole well shaken, and the volume made up to 100 c.c. The solution is filtered and 25 c.c. of the filtrate are neutralised to phenolphthalein (1 drop used), then 20 c.c. $\frac{N}{10}$ iodine and 30 c.c. of $\frac{N}{10}$ soda are added; after twenty minutes 4 c.c. of $\frac{N}{1}$ sulphuric acid are added, and the residual iodine titrated with $\frac{N}{10}$ thiosulphate solution.

The percentage of milk-sugar by weight is calculated thus—

$$\% = \text{c.c. iodine used} \times 0{\cdot}0171 \times \frac{100}{25} \times \frac{100 - (0{\cdot}3 + F \times 0{\cdot}111)}{100} \times \frac{100}{\text{wt. of milk}}.$$

If desired 10 c.c. of milk may be measured and the weight of this taken as 10 × sp. gr.

TABLE OF LOGARITHMS.

	0	1	2	3	4	5	6	7	8	9	1	2	3	4	5	6	7	8	9
10	0000	0043	0086	0128	0170	0212	0253	0294	0334	0374	4	8	12	17	21	25	29	33	37
11	0414	0453	0492	0531	0569	0607	0645	0682	0719	0755	4	8	11	15	19	23	26	30	34
12	0792	0828	0864	0899	0934	0969	1004	1038	1072	1106	3	7	10	14	17	21	24	28	31
13	1139	1173	1206	1239	1271	1303	1335	1367	1399	1430	3	6	10	13	16	19	23	26	29
14	1461	1492	1523	1553	1584	1614	1644	1673	1703	1732	3	6	9	12	15	18	21	24	27
15	1761	1790	1818	1847	1875	1903	1931	1959	1987	2014	3	6	8	11	14	17	20	22	25
16	2041	2068	2095	2122	2148	2175	2201	2227	2253	2279	3	5	8	11	13	16	18	21	24
17	2304	2330	2355	2380	2405	2430	2455	2480	2504	2529	2	5	7	10	12	15	17	20	22
18	2553	2577	2601	2625	2648	2672	2695	2718	2742	2765	2	5	7	9	12	14	16	19	21
19	2788	2810	2833	2856	2878	2900	2923	2945	2967	2989	2	4	7	9	11	13	16	18	20
20	3010	3032	3054	3075	3096	3118	3139	3160	3181	3201	2	4	6	8	11	13	15	17	19
21	3222	3243	3263	3284	3304	3324	3345	3365	3385	3404	2	4	6	8	10	12	14	16	18
22	3424	3444	3464	3483	3502	3522	3541	3560	3579	3598	2	4	6	8	10	12	14	15	17
23	3617	3636	3655	3674	3692	3711	3729	3747	3766	3784	2	4	6	7	9	11	13	15	17
24	3802	3820	3838	3856	3874	3892	3909	3927	3945	3962	2	4	5	7	9	11	12	14	16
25	3979	3997	4014	4031	4048	4065	4082	4099	4116	4133	2	3	5	7	9	10	12	14	15
26	4150	4166	4183	4200	4216	4232	4249	4265	4281	4298	2	3	5	7	8	10	11	13	15
27	4314	4330	4346	4362	4378	4393	4409	4425	4440	4456	2	3	5	6	8	9	11	13	14
28	4472	4487	4502	4518	4533	4548	4564	4579	4594	4609	2	3	5	6	8	9	11	12	14
29	4624	4639	4654	4669	4683	4698	4713	4728	4742	4757	1	3	4	6	7	9	10	12	13
30	4771	4786	4800	4814	4829	4843	4857	4871	4886	4900	1	3	4	6	7	9	10	11	13
31	4914	4928	4942	4955	4969	4983	4997	5011	5024	5038	1	3	4	6	7	8	10	11	12
32	5051	5065	5079	5092	5105	5119	5132	5145	5159	5172	1	3	4	5	7	8	9	11	12
33	5185	5198	5211	5224	5237	5250	5263	5276	5289	5302	1	3	4	5	6	8	9	10	12
34	5315	5328	5340	5353	5366	5378	5391	5403	5416	5428	1	3	4	5	6	8	9	10	11
35	5441	5453	5465	5478	5490	5502	5514	5527	5539	5551	1	2	4	5	6	7	9	10	11
36	5563	5575	5587	5599	5611	5623	5635	5647	5658	5670	1	2	4	5	6	7	8	10	11
37	5682	5694	5705	5717	5729	5740	5752	5763	5775	5786	1	2	3	5	6	7	8	9	10
38	5798	5809	5821	5832	5843	5855	5866	5877	5888	5899	1	2	3	5	6	7	8	9	10
39	5911	5922	5933	5944	5955	5966	5977	5988	5999	6010	1	2	3	4	5	7	8	9	10
40	6021	6031	6042	6053	6064	6075	6085	6096	6107	6117	1	2	3	4	5	6	8	9	10
41	6128	6138	6149	6160	6170	6180	6191	6201	6212	6222	1	2	3	4	5	6	7	8	9
42	6232	6243	6253	6263	6274	6284	6294	6304	6314	6325	1	2	3	4	5	6	7	8	9
43	6335	6345	6355	6365	6375	6385	6395	6405	6415	6425	1	2	3	4	5	6	7	8	9
44	6435	6444	6454	6464	6474	6484	6493	6503	6513	6522	1	2	3	4	5	6	7	8	9
45	6532	6542	6551	6561	6571	6580	6590	6599	6609	6618	1	2	3	4	5	6	7	8	9
46	6628	6637	6646	6656	6665	6675	6684	6693	6702	6712	1	2	3	4	5	6	7	7	8
47	6721	6730	6739	6749	6758	6767	6776	6785	6794	6803	1	2	3	4	5	5	6	7	8
48	6812	6821	6830	6839	6848	6857	6866	6875	6884	6893	1	2	3	4	4	5	6	7	8
49	6902	6911	6920	6928	6937	6946	6955	6964	6972	6981	1	2	3	4	4	5	6	7	8
50	6990	6998	7007	7016	7024	7033	7042	7050	7059	7067	1	2	3	3	4	5	6	7	8
51	7076	7084	7093	7101	7110	7118	7126	7135	7143	7152	1	2	3	3	4	5	6	7	8
52	7160	7168	7177	7185	7193	7202	7210	7218	7226	7235	1	2	2	3	4	5	6	7	7
53	7243	7251	7259	7267	7275	7284	7292	7300	7308	7316	1	2	2	3	4	5	6	6	7
54	7324	7332	7340	7348	7356	7364	7372	7380	7388	7396	1	2	2	3	4	5	6	6	7

TABLE OF LOGARITHMS.—*Continued.*

	0	1	2	3	4	5	6	7	8	9	1	2	3	4	5	6	7	8	9
55	7404	7412	7419	7427	7435	7443	7451	7459	7466	7474	1	2	2	3	4	5	5	6	7
56	7482	7490	7497	7505	7513	7520	7528	7536	7543	7551	1	2	2	3	4	5	5	6	7
57	7559	7566	7574	7582	7589	7597	7604	7612	7619	7627	1	2	2	3	4	5	5	6	7
58	7634	7642	7649	7657	7664	7672	7679	7686	7694	7701	1	1	2	3	4	4	5	6	7
59	7709	7716	7723	7731	7738	7745	7752	7760	7767	7774	1	1	2	3	4	4	5	6	7
60	7782	7789	7796	7803	7810	7818	7825	7832	7839	7846	1	1	2	3	4	4	5	6	6
61	7853	7860	7868	7875	7882	7889	7896	7903	7910	7917	1	1	2	3	4	4	5	6	6
62	7924	7931	7938	7945	7952	7959	7966	7973	7980	7987	1	1	2	3	3	4	5	6	6
63	7993	8000	8007	8014	8021	8028	8035	8041	8048	8055	1	1	2	3	3	4	5	5	6
64	8062	8069	8075	8082	8089	8096	8102	8109	8116	8122	1	1	2	3	3	4	5	5	6
65	8129	8136	8142	8149	8156	8162	8169	8176	8182	8189	1	1	2	3	3	4	5	5	6
66	8195	8202	8209	8215	8222	8228	8235	8241	8248	8254	1	1	2	3	3	4	5	5	6
67	8261	8267	8274	8280	8287	8293	8299	8306	8312	8319	1	1	2	3	3	4	5	5	6
68	8325	8331	8338	8344	8351	8357	8363	8370	8376	8382	1	1	2	3	3	4	4	5	6
69	8388	8395	8401	8407	8414	8420	8426	8432	8439	8445	1	1	2	2	3	4	4	5	6
70	8451	8457	8463	8470	8476	8482	8488	8494	8500	8506	1	1	2	2	3	4	4	5	6
71	8513	8519	8525	8531	8537	8543	8549	8555	8561	8567	1	1	2	2	3	4	4	5	5
72	8573	8579	8585	8591	8597	8603	8609	8615	8621	8627	1	1	2	2	3	4	4	5	5
73	8633	8639	8645	8651	8657	8663	8669	8675	8681	8686	1	1	2	2	3	4	4	5	5
74	8692	8698	8704	8710	8716	8722	8727	8733	8739	8745	1	1	2	2	3	4	4	5	5
75	8751	8756	8762	8768	8774	8779	8785	8791	8797	8802	1	1	2	2	3	3	4	5	5
76	8808	8814	8820	8825	8831	8837	8842	8848	8854	8859	1	1	2	2	3	3	4	5	5
77	8865	8871	8876	8882	8887	8893	8899	8904	8910	8915	1	1	2	2	3	3	4	4	5
78	8921	8927	8932	8938	8943	8949	8954	8960	8965	8971	1	1	2	2	3	3	4	4	5
79	8976	8982	8987	8993	8998	9004	9009	9015	9020	9025	1	1	2	2	3	3	4	4	5
80	9031	9036	9042	9047	9053	9058	9063	9069	9074	9079	1	1	2	2	3	3	4	4	5
81	9085	9090	9096	9101	9106	9112	9117	9122	9128	9133	1	1	2	2	3	3	4	4	5
82	9138	9143	9149	9154	9159	9165	9170	9175	9180	9186	1	1	2	2	3	3	4	4	5
83	9191	9196	9201	9206	9212	9217	9222	9227	9232	9238	1	1	2	2	3	3	4	4	5
84	9243	9248	9253	9258	9263	9269	9274	9279	9284	9289	1	1	2	2	3	3	4	4	5
85	9294	9299	9304	9309	9315	9320	9325	9330	9335	9340	1	1	2	2	3	3	4	4	5
86	9345	9350	9355	9360	9365	9370	9375	9380	9385	9390	1	1	2	2	3	3	4	4	5
87	9395	9400	9405	9410	9415	9420	9425	9430	9435	9440	0	1	1	2	2	3	3	4	4
88	9445	9450	9455	9460	9465	9469	9474	9479	9484	9489	0	1	1	2	2	3	3	4	4
89	9494	9499	9504	9509	9513	9518	9523	9528	9533	9538	0	1	1	2	2	3	3	4	4
90	9542	9547	9552	9557	9562	9566	9571	9576	9581	9586	0	1	1	2	2	3	3	4	4
91	9590	9595	9600	9605	9609	9614	9619	9624	9628	9633	0	1	1	2	2	3	3	4	4
92	9638	9643	9647	9652	9657	9661	9666	9671	9675	9680	0	1	1	2	2	3	3	4	4
93	9685	9689	9694	9699	9703	9708	9713	9717	9722	9727	0	1	1	2	2	3	3	4	4
94	9731	9736	9741	9745	9750	9754	9759	9763	9768	9773	0	1	1	2	2	3	3	4	4
95	9777	9782	9786	9791	9795	9800	9805	9809	9814	9818	0	1	1	2	2	3	3	4	4
96	9823	9827	9832	9836	9841	9845	9850	9854	9859	9863	0	1	1	2	2	3	3	4	4
97	9868	9872	9877	9881	9886	9890	9894	9899	9903	9908	0	1	1	2	2	3	3	4	4
98	9912	9917	9921	9926	9930	9934	9939	9943	9948	9952	0	1	1	2	2	3	3	4	4
99	9956	9961	9965	9969	9974	9978	9983	9987	9991	9996	0	1	1	2	2	3	3	3	4

TABLE OF ANTILOGARITHMS.

	0	1	2	3	4	5	6	7	8	9	1	2	3	4	5	6	7	8	9
·00	1000	1002	1005	1007	1009	1012	1014	1016	1019	1021	0	0	1	1	1	1	2	2	2
·01	1023	1026	1028	1030	1033	1035	1038	1040	1042	1045	0	0	1	1	1	1	2	2	2
·02	1047	1050	1052	1054	1057	1059	1062	1064	1067	1069	0	0	1	1	1	1	2	2	2
·03	1072	1074	1076	1079	1081	1084	1086	1089	1091	1094	0	0	1	1	1	1	2	2	2
·04	1096	1099	1102	1104	1107	1109	1112	1114	1117	1119	0	1	1	1	1	2	2	2	2
·05	1122	1125	1127	1130	1132	1135	1138	1140	1143	1146	0	1	1	1	1	2	2	2	2
·06	1148	1151	1153	1156	1159	1161	1164	1167	1169	1172	0	1	1	1	1	2	2	2	2
·07	1175	1178	1180	1183	1186	1189	1191	1194	1197	1199	0	1	1	1	1	2	2	2	2
·08	1202	1205	1208	1211	1213	1216	1219	1222	1225	1227	0	1	1	1	1	2	2	2	3
·09	1230	1233	1236	1239	1242	1245	1247	1250	1253	1256	0	1	1	1	1	2	2	2	3
·10	1259	1262	1265	1268	1271	1274	1276	1279	1282	1285	0	1	1	1	1	2	2	2	3
·11	1288	1291	1294	1297	1300	1303	1306	1309	1312	1315	0	1	1	1	2	2	2	2	3
·12	1318	1321	1324	1327	1330	1334	1337	1340	1343	1346	0	1	1	1	2	2	2	2	3
·13	1349	1352	1355	1358	1361	1365	1368	1371	1374	1377	0	1	1	1	2	2	2	3	3
·14	1380	1384	1387	1390	1393	1396	1400	1403	1406	1409	0	1	1	1	2	2	2	3	3
·15	1413	1416	1419	1422	1426	1429	1432	1435	1439	1442	0	1	1	1	2	2	2	3	3
·16	1445	1449	1452	1455	1459	1462	1466	1469	1472	1476	0	1	1	1	2	2	2	3	3
·17	1479	1483	1486	1489	1493	1496	1500	1503	1507	1510	0	1	1	1	2	2	2	3	3
·18	1514	1517	1521	1524	1528	1531	1535	1538	1542	1545	0	1	1	1	2	2	2	3	3
·19	1549	1552	1556	1560	1563	1567	1570	1574	1578	1581	0	1	1	1	2	2	3	3	3
·20	1585	1589	1592	1596	1600	1603	1607	1611	1614	1618	0	1	1	1	2	2	3	3	3
·21	1622	1626	1629	1633	1637	1641	1644	1648	1652	1656	0	1	1	2	2	2	3	3	3
·22	1660	1663	1667	1671	1675	1679	1683	1687	1690	1694	0	1	1	2	2	2	3	3	3
·23	1698	1702	1706	1710	1714	1718	1722	1726	1730	1734	0	1	1	2	2	2	3	3	4
·24	1738	1742	1746	1750	1754	1758	1762	1766	1770	1774	0	1	1	2	2	2	3	3	4
·25	1778	1782	1786	1791	1795	1799	1803	1807	1811	1816	0	1	1	2	2	2	3	3	4
·26	1820	1824	1828	1832	1837	1841	1845	1849	1854	1858	0	1	1	2	2	3	3	3	4
·27	1862	1866	1871	1875	1879	1884	1888	1892	1897	1901	0	1	1	2	2	3	3	3	4
·28	1905	1910	1914	1919	1923	1928	1932	1936	1941	1945	0	1	1	2	2	3	3	4	4
·29	1950	1954	1959	1963	1968	1972	1977	1982	1986	1991	0	1	1	2	2	3	3	4	4
·30	1995	2000	2004	2009	2014	2018	2023	2028	2032	2037	0	1	1	2	2	3	3	4	4
·31	2042	2046	2051	2056	2061	2065	2070	2075	2080	2084	0	1	1	2	2	3	3	4	4
·32	2089	2094	2099	2104	2109	2113	2118	2123	2128	2133	0	1	1	2	2	3	3	4	4
·33	2138	2143	2148	2153	2158	2163	2168	2173	2178	2183	0	1	1	2	2	3	3	4	4
·34	2188	2193	2198	2203	2208	2213	2218	2223	2228	2234	1	1	2	2	3	3	4	4	5
·35	2239	2244	2249	2254	2259	2265	2270	2275	2280	2286	1	1	2	2	3	3	4	4	5
·36	2291	2296	2301	2307	2312	2317	2323	2328	2333	2339	1	1	2	2	3	3	4	4	5
·37	2344	2350	2355	2360	2366	2371	2377	2382	2388	2393	1	1	2	2	3	3	4	4	5
·38	2399	2404	2410	2415	2421	2427	2432	2438	2443	2449	1	1	2	2	3	3	4	4	5
·39	2455	2460	2466	2472	2477	2483	2489	2495	2500	2506	1	1	2	2	3	3	4	5	5
·40	2512	2518	2523	2529	2535	2541	2547	2553	2559	2564	1	1	2	2	3	4	4	5	5
·41	2570	2576	2582	2588	2594	2600	2606	2612	2618	2624	1	1	2	2	3	4	4	5	5
·42	2630	2636	2642	2649	2655	2661	2667	2673	2679	2685	1	1	2	2	3	4	4	5	6
·43	2692	2698	2704	2710	2716	2723	2729	2735	2742	2748	1	1	2	3	3	4	4	5	6
·44	2754	2761	2767	2773	2780	2786	2793	2799	2805	2812	1	1	2	3	3	4	4	5	6
·45	2818	2825	2831	2838	2844	2851	2858	2864	2871	2877	1	1	2	3	3	4	5	5	6
·46	2884	2891	2897	2904	2911	2917	2924	2931	2938	2944	1	1	2	3	3	4	5	5	6
·47	2951	2958	2965	2972	2979	2985	2992	2999	3006	3013	1	1	2	3	3	4	5	5	6
·48	3020	3027	3034	3041	3048	3055	3062	3069	3076	3083	1	1	2	3	4	4	5	6	6
·49	3090	3097	3105	3112	3119	3126	3133	3141	3148	3155	1	1	2	3	4	4	5	6	6

TABLE OF ANTILOGARITHMS.—*Continued.*

	0	1	2	3	4	5	6	7	8	9	1	2	3	4	5	6	7	8	9
·50	3162	3170	3177	3184	3192	3199	3206	3214	3221	3228	1	1	2	3	4	4	5	6	7
·51	3236	3243	3251	3258	3266	3273	3281	3289	3296	3304	1	2	2	3	4	5	5	6	7
·52	3311	3319	3327	3334	3342	3350	3357	3365	3373	3381	1	2	2	3	4	5	5	6	7
·53	3388	3396	3404	3412	3420	3428	3436	3443	3451	3459	1	2	2	3	4	5	6	6	7
·54	3467	3475	3483	3491	3499	3508	3516	3524	3532	3540	1	2	2	3	4	5	6	6	7
·55	3548	3556	3565	3573	3581	3589	3597	3606	3614	3622	1	2	2	3	4	5	6	7	7
·56	3631	3639	3648	3656	3664	3673	3681	3690	3698	3707	1	2	3	3	4	5	6	7	8
·57	3715	3724	3733	3741	3750	3758	3767	3776	3784	3793	1	2	3	3	4	5	6	7	8
·58	3802	3811	3819	3828	3837	3846	3855	3864	3873	3882	1	2	3	4	4	5	6	7	8
·59	3890	3899	3908	3917	3926	3936	3945	3954	3963	3972	1	2	3	4	5	5	6	7	8
·60	3981	3990	3999	4009	4018	4027	4036	4046	4055	4064	1	2	3	4	5	6	6	7	8
·61	4074	4083	4093	4102	4111	4121	4130	4140	4150	4159	1	2	3	4	5	6	7	8	9
·62	4169	4178	4188	4198	4207	4217	4227	4236	4246	4256	1	2	3	4	5	6	7	8	9
·63	4266	4276	4285	4295	4305	4315	4325	4335	4345	4355	1	2	3	4	5	6	7	8	9
·64	4365	4375	4385	4395	4406	4416	4426	4436	4446	4457	1	2	3	4	5	6	7	8	9
·65	4467	4477	4487	4498	4508	4519	4529	4539	4550	4560	1	2	3	4	5	6	7	8	9
·66	4571	4581	4592	4603	4613	4624	4634	4645	4656	4567	1	2	3	4	5	6	7	9	10
·67	4677	4688	4699	4710	4721	4732	4742	4753	4764	4775	1	2	3	4	5	7	8	9	10
·68	4786	4797	4808	4819	4831	4842	4853	4864	4875	4887	1	2	3	4	6	7	8	9	10
·69	4898	4909	4920	4932	4943	4955	4966	4977	4989	5000	1	2	3	5	6	7	8	9	10
·70	5012	5023	5035	5047	5058	5070	5082	5093	5105	5117	1	2	4	5	6	7	8	9	11
·71	5129	5140	5152	5164	5176	5188	5200	5212	5224	5236	1	2	4	5	6	7	8	10	11
·72	5248	5260	5272	5284	5297	5309	5321	5333	5346	5358	1	2	4	5	6	7	9	10	11
·73	5370	5383	5395	5408	5420	5433	5445	5458	5470	5483	1	3	4	5	6	8	9	10	11
·74	5495	5508	5521	5534	5546	5559	5572	5585	5598	5610	1	3	4	5	0	8	9	10	12
·75	5623	5636	5649	5662	5675	5689	5702	5715	5728	5741	1	3	4	5	7	8	9	10	12
·76	5754	5768	5781	5794	5808	5821	5834	5848	5861	5875	1	3	4	5	7	8	9	11	12
·77	5888	5902	5916	5929	5943	5957	5970	5984	5998	6012	1	3	4	5	7	8	10	11	12
·78	6026	6039	6053	6067	6081	6095	6109	6124	6138	6152	1	3	4	6	7	8	10	11	13
·79	6166	6180	6194	6209	6223	6237	6252	6266	6281	6295	1	3	4	6	7	9	10	11	13
·80	6310	6324	6339	6353	6368	6383	6397	6412	6427	6442	1	3	4	6	7	9	10	12	13
·81	6457	6471	6486	6501	6516	6531	6546	6561	6577	6592	2	3	5	6	8	9	11	12	14
·82	6607	6622	6637	6653	6668	6683	6699	6714	6730	6745	2	3	5	6	8	9	11	12	14
·83	6761	6776	6792	6808	6823	6839	6855	6871	6887	6902	2	3	5	6	8	9	11	13	14
·84	6918	6934	6950	6966	6982	6998	7015	7031	7047	7063	2	3	5	6	8	10	11	13	15
·85	7079	7096	7112	7129	7145	7161	7178	7194	7211	7228	2	3	5	7	8	10	12	13	15
·86	7244	7261	7278	7295	7311	7328	7345	7362	7379	7396	2	3	5	7	8	10	12	13	15
·87	7413	7430	7447	7464	7482	7499	7516	7534	7551	7568	2	3	5	7	9	10	12	14	16
·88	7586	7603	7621	7638	7656	7674	7691	7709	7727	7745	2	4	5	7	9	11	12	14	16
·89	7762	7780	7798	7816	7834	7852	7870	7889	7907	7925	2	4	5	7	9	11	13	14	16
·90	7943	7962	7980	7998	8017	8035	8054	8072	8891	8110	2	4	6	7	9	11	13	15	17
·91	8128	8147	8166	8185	8204	8222	8241	8260	8279	8299	2	4	6	8	9	11	13	15	17
·92	8318	8337	8356	8375	8395	8414	8433	8453	8472	8492	2	4	6	8	10	12	14	15	17
·93	8511	8531	8551	8570	8590	8610	8630	8650	8670	8690	2	4	6	8	10	12	14	16	18
·94	8710	8730	8750	8770	8790	8810	8831	8851	8872	8892	2	4	6	8	10	12	14	16	18
·95	8913	8933	8954	8974	8995	9016	9036	9057	9078	9099	2	4	6	8	10	12	15	17	19
·96	9120	9141	9162	9183	9204	9226	9247	9268	9290	9311	2	4	6	8	11	13	15	17	19
·97	9333	9354	9376	9397	9419	9441	9462	9484	9506	9528	2	4	7	9	11	13	15	17	20
·98	9550	9572	9594	9616	9638	9661	9683	9705	9727	9750	2	4	7	9	11	13	16	18	20
·99	9772	9795	9817	9840	9863	9886	9908	9931	9954	9977	2	5	7	9	11	14	16	18	20

INDEX.

M

N

O

P

R

S

BAIRD & TATLOCK (LONDON) LTD.'S LIST OF APPARATUS AND CHEMICALS REQUIRED FOR MILK ANALYSIS, &c.

As mentioned in the "Laboratory Book of Dairy Analysis,"
by H. DROOP RICHMOND, F.I.C.

ALL the apparatus detailed below are required for a full equipment; the quantity will vary according to requirements :—

THE ANALYSIS OF MILK.

Sampling.

Page.	Item No.	
5	1	Sample bottles, oval shape, with corks, 5 ozs.
6	2	Beakers, with spout, No. 5.
6	3	Brush, fine wire, for mixing.

Specific Gravity.

Page.	Item No.	
6	4	Lactometer, ordinary pattern, paper scale.
7	5	Lactometer, Veith's pattern, 25-35.
7	6	Lactometer, Soxlhet's pattern, about 13 inches long.
7	7	Lactometer jar, zinc, 4 inches by $1\frac{3}{4}$ inches, with spout.
8	8	Thermo-lactometer, Quevenne's.
9	9	Sprengel tube, 10 c.c.
9	9*a*	Water oven.
10	10	Beaker flasks, 200 c.c.

Estimation of Total Solids.

Page.	Item No.	
11	11	Platinum basin, flat, shallow, with spout, $2\frac{3}{4}$ inches.
11	12	Porcelain capsule, glazed all over.
	12*a*	Tantalum basin.
	12*b*	Silica basin.
11	13	Pipette for milk, 5 c.c., with mark.
11	14	Pipette to hold 5 grammes of milk.
11	15	Water-bath to take 6 porcelain capsules.
12	16	Desiccator, Scheibler's pattern, 6 inches.
12	17	Balance, Bunge's 200 grammes, sensibility to $\frac{1}{10}$th mgm. on plate-glass sole.
12	18	Set of gilt weights for above 50 to ·001 grammes.

Estimation of Ash, etc.

Page.	*Item No.*	
12	19	Muffle furnace, No. 461.
12	20	Fireclay muffle for ditto.
12	21	Platinum wire, 6 inches long.
12	22	Tripod, triangular, 8 inches high.
12	23	Bunsen burner, ½ inch.
		Argand burner.
12	24	India-rubber tubing for above, heavy walls.
12	25	Pipe-clay triangle, Clowes' improved form.
		Triangle silica.
13	26	Filter-paper, C.S. and S. 589, black band.
15	27	Porcelain basin, round-bottomed, with spout, diameter 3½ inches.

Estimation of Acidity.

16 28 Acidimeter for testing acidity, consisting of special burette, stand, and dropping bottle.

Estimation of Fat.

17 29 Gerber's butyrometer for 4 samples, with accessories as follows :—

4 bottles ;
1 pipette, 3 c.c. for cream ;
1 „ 10 c.c. 100 divs. for water ;
1 „ 11 c.c. for milk.
1 „ 10 c.c. for acid ;
1 „ 1 c.c. for amyl alcohol.

17 30 Leffman-Beam centrifugal machine, for 4 samples, with accessories as follows :—

4 bottles ;
1 pipette, 3 c.c. ;
1 „ 9 c.c. ;
1 „ 15 c.c.

20 31 Richmond's shaking stand for 8 Gerber's tubes.

25 32 Pair of tared tin dishes, 2 inches by 1⅝ inches for weighing cream.

Gravimetric Estimation of Fat.

Page.	*Item No.*	
29	34	Werner-Schmid tube.
28	35	Fat extraction thimble.
28	36	Glass mortar, 3 inches diameter, with pestle.
35	37	Funnel, $2\frac{1}{2}$ inches.
35	38	Filter-paper, C.S. and S 595, to suit.
35	39	Beakers, No. 3, with spout.
35	40	6 glass stirring rods, 6 inches long.
35	41	Conical flasks, 6 ozs.
	41*a*	Stoppered cylinders.
	41*b*	Wash-bottle tubes.
29	42	Soxhlet's Fat Extraction Apparatus as follows :— 1 water-bath for 3 apparatus on stand ; 1 set Bunsen's for ditto ; 3 fat extractors, 60 grammes ; 3 flasks with short necks, 4 ozs. ; 3 condensers with india-rubber caps ; 1 stand with three-armed clamp for supporting condensers.
30	43	Stokes tube.
30	44	Burette, 100 c.c. $\frac{1}{10}$ths, with stopcock. Mahogany stand for ditto.

Estimation of Milk-Sugar.

32	45	Mitscherlich's polariscope, with 100 and 200 mm. tube and burner.
32	46	Hard glass tubes, 10 × 1 c.m., with one end drawn down.
31	47	Flasks, with mark 100 c.c., unstoppered.
33	48	Policemen, with india-rubber tops, for stirring.
33	49	Filter pump glass.
33	50	Glass tubing, assorted.
33	51	Wash-bottles, fitted with cork and tubes, 1 litre.

Estimation of Proteins.

34	52	Gooch's porcelain crucible.
34	53	Asbestos fibre, for use with above.

Estimation of Casein and Albumin.

35	54	Porcelain crucibles, No. 1, with lids.
35	55	Crucible tongs, gun-metal, with platinum tips.

Estimation of Nitrogen.

Page.	*Item No.*	
35	56	Kjeldahl's apparatus, with litre copper flask, Liebig's condenser, etc.
35	57	Flask, 200 c.c., hard glass, round bottom.
40	58	Burette, fitted with soda lime tube, and bottle with soda lime tube to hold strontia solution.
41	58*a*	Richmond's milk slide rule.
44	58*c*	Richmond improved slide rule.
46	58*d*	Collins' milk scale.

Analysis of Liquid Milk-Products.

47	59	Pipette, 5 c.c., short form with india-rubber tube at top.
48	60	Flask, with side tube, 250 c.c.
48	61	Liebig's condenser, for use with ditto.
41	61*a*	Lobeck's catalase tubes.
49	62	Nessler tubes, marked at 50 c.c.

The Application of Analysis to the Solutions of Problems.

60	62*a*	Separating funnel for salicylic and benzoic acid.
60	63	Test-tubes, $6 \times \frac{5}{8}$ inch.
62	64	High-speed centrifuge.
64	65	Microscope, with $\frac{2}{3}$ inch and $\frac{1}{6}$ inch objective, one eyepiece and substage condenser and polariscope fitted.
64	65*a*	Test-tubes, $2 \times \frac{1}{4}$ inch.
	66	2 burettes, 50 c.c. $\frac{1}{10}$ths.
	67	Stand, double, for ditto.
	68	6 dropping bottles, with grooved stopper, 30 c.c.

THE ANALYSIS OF BUTTER.

Estimation of Water, etc.

72	69	Sand-bath, 6 inches.
72	70	1 round tripod for ditto.
72	71	1 Bunsen burner for ditto.
72	72	Porcelain basins, No. 1.
72	73	Glass rod for ditto.
74	74	Glass plates, $2\frac{1}{2} \times 2\frac{1}{2}$ inches.
74	75	Set of pipettes, with mark 5, 10, 20, and 25 c.c.

Estimation of the Volatile Fatty Acids.

Page.	*Item No.*	
77	76	Caustic soda apparatus.
78	77	Polenske apparatus.
79	78	Flask, CO_2, 6 ozs.
79	79	Liebig's condenser for ditto.
79	80	Porcelain basin, 6 inches diameter.
84	81	Microscope slides, 3 × 1 inch.
84	82	Microscope cover glasses, $\frac{5}{8}$ inch square, No. 2.
85	83	Bottles, 4 ozs. N.M., flat-stoppered.

Determination of Refractive Index.

86	84	Zeiss' butyro-refractometer.

THE ANALYSIS OF CHEESE.

96	85	Porcelain capsules, glazed all over.
97	86	Flask, 250 c.c. with mark, stoppered.
97	87	Separators, 500 c.c. cylindrical, with stopper and stopcock.
98	88	Hot-air bath, 7 × 7 inches, on stand.
98	89	Thermometer, engraved up to 110° C.
99	90	Incubator, B.T.L., slag wool lined, with capsule regulating at 37° C., complete with thermometer.
99	91	Pair watch-glasses with binder.
99	92	100 c.c. flask with mark.

CHEMICALS AND STANDARD SOLUTIONS.

THE quantities detailed below are necessary for a large laboratory. They may be reduced in proportion when only a small outfit is required :—

5 lbs. acetic acid 100 per cent.
1 lb. acetone.
4 ozs. phenolphthalein indicator, 0·5 per cent.
1 litre sulphuric acid $\frac{N}{10}$ solution.
4 ozs. potassium chromate indicator.
1 litre silver nitrate $\frac{N}{10}$ solution.
1 w. qt. hydrochloric acid, pure.
1 w. qt. ammonia, ·880.

2 lbs. Schering's formalin.
½ lb. ammonium oxalate, pure.
½ lb. ammonium carbonate, pure.
½ lb. magnesia mixture.
1 litre caustic soda $\frac{N}{10}$ solution.
1 lb. Kieselguhr special for filtering.
1 w. qt. amylic alcohol } for Leffmann-Beam or Gerber
1 carboy sulphuric acid } process.
1 lb. soda lime.
1 w. qt. alcohol.
1 w. qt. petroleum ether.
5 lbs. caustic soda, pure. }
1 w. qt. sulphuric acid free N. }
1 lb. mercury redistilled. } for Kjeldahl's
2 lbs. potassium bisulphate. } process.
1 lb. sodium sulphide pure reagent. }
4 ozs. cochineal indicator. }
1 lb. phosphotungstic acid.
2 lbs. magnesium sulphate, pure.
1 w. qt. nitric acid, pure.
2 lbs. sodium carbonate.
1 lb. barium chloride.
1 w. qt. ether ·720 meth.
1 tube rennet tablets.
1 lb. beeswax.
1 roll each blue and red litmus paper and turmeric paper.
½ lb. ferric chloride.
1 oz. para-phenylenediamine.
1 oz. ortol.
1 oz. resorcine.
2 lbs. sodium bicarbonate.
½ lb. potassium ferrocyanide, pure.
½ lb. iodine re-sublimed.
1 lb. potassium iodide.
2 lbs. hydrogen peroxide, 20 vol.
2 oz. picric acid.
½ oz. silver nitrate.
1 lb. calcium chloride, dry.
1 lb. chloroform.

2 lbs. Rochelle salt.
1 oz. furfural.
2 lbs. copper sulphate.
1 lb. mercuric chloride.
1 lb. magnesium chloride.
5 lbs. caustic potash.
¼ lb. strontium hydrate.
1 lb. sodium thiosulphate.
1 oz. salicylic acid.
½ litre Nessler's reagent.
2 lbs. glycerine.
½ lb. bromine.
⅛ lb. pumice.